JAN − 3 1995

D1439140

THE WEST
PACIFIC RIM
AN INTRODUCTION

RUPERT HODDER

The West Pacific Rim

To San

The West Pacific Rim

An Introduction

Rupert Hodder

Belhaven Press
London

Co-published in the Americas by Halsted Press,
an imprint of John Wiley & Sons, Inc., New York

Belhaven Press
(a division of Pinter Publishers)
25 Floral Street, Covent Garden, London, WC2E 9DS, United Kingdom

First published in 1992

Co-published in the Americas by Halsted Press, an imprint of John Wiley & Sons, Inc., 605 Third Avenue, New York, NY 10158–0012, USA

Rupert Hodder is hereby identified as the author of this work as provided under Section 77 of the Copyright, Designs and Patents Act, 1988.

British Library Cataloguing in Publication Data

A CIP catalogue record for this book is available from the British Library.

ISBN 1 85293 210 4 (hb)
   1 85293 211 2 (pbk)

Library of Congress Cataloging-in-Publication Data

Hodder, Rupert.
  The West Pacific rim: an introduction / Rupert Hodder.
    p.   cm.
  Includes bibliographical references and index.
  ISBN 1-85293-210-4. — ISBN 1-85293-211-2 (pbk.). — ISBN 0-470-21963-7 (Americas only). — ISBN 0-470-21964-5 (pbk. : Americas only)
    1. East Asia—Economic conditions.   2. East Asia—Economic policy. 3. Pacific Area—Economic conditions.   4. Pacific Area—Economic policy.   I. Title.
HC460.5.H6   1992
330.951—dc20   92–27957
                                          CIP

ISBN 0-470-21963-7 (hb in the Americas only)
     0-470-21964-5 (pbk in the Americas only)

Typeset by Florencetype Ltd, Kewstoke, Avon
Printed and bound in Great Britain by
Biddles Ltd of Guildford and King's Lynn

Contents

List of figures vi
List of tables vii
Preface viii

Introduction 1
1 Patterns and resources 6
2 Historical perspectives and the colonial legacy 23
3 Culture and development in the West Pacific Rim 34
4 The Chinese diaspora 43
5 Exchange and domestic trade 56
6 International trade and investment 65
7 Economic strategies and government intervention 79
8 Japan and the four NICs 90
9 China since 1978 107
10 The Southeast Asian sector 118
11 Pacific concerns 129

Conclusion 136
References and further reading 142
Index 149

List of figures

Figure I.1	The West Pacific Rim	2
Figure 1.1	Income inequality and growth of GDP per person 1965–89	9
Figure 6.1	Estimated percentage annual growth in real exports, selected regions 1965–89	69
Figure 6.2	United States: trade balances with the European Community, the NICs and Japan, 1986–91	74
Figure 7.1	Prybyla's 'Bird in cage' model of adjustment and reform	82
Figure 9.1	China: trade balances and changes in consumer prices and industrial production, 1987–91	116
Figure C.1	Periods during which output per person doubled in selected countries	137

List of tables

Table 1.1	Population and GNP (per capita) and growth	7
Table 1.2	Structure of production, 1965 and 1989	13
Table 1.3	Selected demographic data, 1989	16
Table 1.4	Selected health data	19
Table 1.5	Selected education data, 1988	20
Table 4.1	Numbers of Chinese in Southeast Asia	44
Table 6.1	Growth of merchandise trade, 1980–89	68
Table 6.2	Direction of trade, 1989	71
Table 6.3	Structure of merchandise imports, 1989	73
Table 6.4	Structure of merchandise exports, 1989	73

Preface

The purpose of this book is to provide a brief and highly selective introduction to the West Pacific Rim, a part of the world to which increasing attention is now being directed. It contains several countries – notably Japan and the four 'Little Tigers' of Singapore, Hong Kong, South Korea and Taiwan – that have achieved remarkable levels of economic prosperity since 1945, when all of them were suffering from agrarian or urban poverty or from the destruction of their economies. Most of the other countries in the region, even including China, are also beginning to prosper significantly, especially compared with most other parts of the developing world. Why is this? And what are the implications of the achievements of West Pacific Rim countries, not only for the wider Pacific world but also for the entire global economy? If, as many predict, Japan achieves world economic hegemony, and if China, with its large population, resources and market, ever fulfils its enormous economic potential, what effect will this have on the world economy and the balance of power? All these are questions of immediate practical as well as academic interest.

In writing this short introductory book on such a vast and diverse area, I have chosen to focus exclusively on trying to answer these questions, articulated more fully in the Introduction, rather than attempting to provide a summary account of the geography, peoples or economies of the various countries in the region. For this reason the discussion is very selective. There is no direct discussion here of the many serious problems – such as urbanisation, housing, agriculture, industry, transportation and the environment – faced in most countries of the West Pacific Rim. While these problems are of course immensely important matters in their own right, and properly form an essential element of the lecture course on part of which this book is based, they have not been considered separately in these pages.

If, on occasions, the reader finds the style rather polemical, this is because this book is designed to stimulate thought and discussion rather than simply to inform. For this reason, too, the discussion and suggested answers to the questions asked in this book are intended to reveal a point of view, or line of argument, which is personal and may well be controversial; however hard one tries, it is impossible to avoid some element of bias or prejudice in covering such a wide range of topics and ideas.

I am grateful to many people, including those undergraduates at the London School of Economics and Political Science who have listened, criticised and argued and so helped me to refine and firm up my own ideas about the West Pacific Rim countries. Professor Robert Bennett and Professor Derek Diamond first gave me the opportunity to develop a course of lectures on the West Pacific Rim and I am grateful to them for their continued support and encouragement. To Frank Leeming, of Leeds University, I will always be indebted for his support in so many ways. My thanks, too, to Jane Pugh, of the Department of Geography at the London School of Economics and Political Science, for her expert drawing of the figures.

Rupert Hodder
London School of Economics and Political Science
January, 1992

Introduction

There are two main reasons for the recent surge of interest in the West Pacific Rim. First, it is now widely suggested that the area is fast becoming part of the new centre of gravity of the world economy. The economic status achieved by so many countries in the region is remarkable and it has become almost fashionable to regard the West Pacific Rim as the most dynamic part of the world. East Asia's share of world GNP (excluding East Europe) increased from 9 per cent in 1962 to 18 per cent in 1988, and it is predicted that the figure will rise to 22 per cent by the year 2000 – by which time over 50 per cent of the world's population will reside in the area (Drysdale, 1988, p. 15; see also Aikman, 1986). By then, too, the sum of the GNPs of Japan, China and the four Newly Industrialised Countries (NICs) will equal that of the United States. Secondly, the West Pacific Rim countries are part of an area of increasing geopolitical and geostrategic significance, both regionally and globally.

The term West Pacific Rim is taken here to apply to the arc of insular, peninsular and littoral countries extending along the West Pacific seaboard[1]. These include, from north to south, the countries of Japan, South Korea, Taiwan, China, Hong Kong, Vietnam, the Philippines, Malaysia, Brunei, Singapore and Indonesia (Figure I.1). In the literature several other terms such as East Asia or the Far East are used to refer to an area roughly the same as that defined in this book as the West Pacific Rim. These terms are used here interchangeably, but the term West Pacific Rim is preferred where the discussion involves the area's Pacific rather than its Asian orientations.

This definition of the West Pacific Rim is of course somewhat arbitrary, and one could argue at length about its precise geographical limits. No doubt a case could be made for including North Korea, Macao, Thailand, Laos, Cambodia, Papua New Guinea or the major Pacific Island groups like Micronesia. However, apart from the fact that I have preferred to write about countries of which I have some experience, the exact regional limits of what I have called the West Pacific Rim

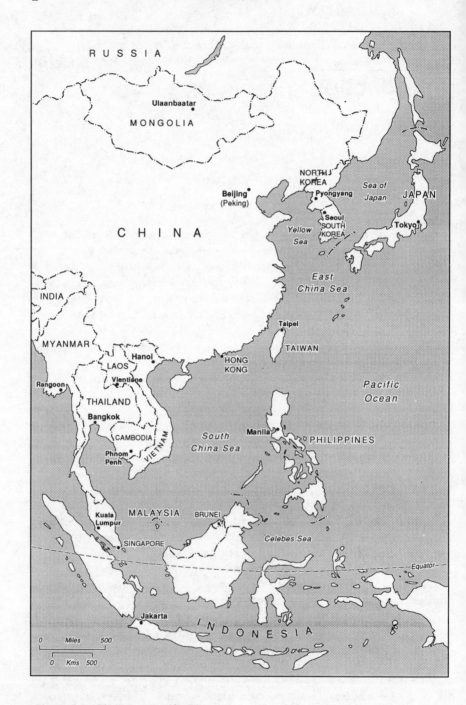

Figure I.1 The West Pacific Rim

– especially in its southern, Southeast Asian sector – have little bearing on the arguments presented in these pages. No attempt has been made to identify any kind of regional identity or unity among the states of the West Pacific Rim; it is too vast, too diverse and too complex to sum up in regional terms. Moreover, the discussion is highly selective in that it is designed solely to assist the understanding of and to suggest answers to a series of questions about this remarkable and fascinating part of the world.

Three groups of questions

Three groups of questions receive particular attention in these pages. First, what explains the remarkable economic success achieved by so many countries in the area? Why are these East Asian countries as a whole economically far more successful than other parts of the Third World? Indeed, the main exceptions to the generally dismal achievements of most developing countries in trying to bring about their own economic development over the last three decades or so are to be found in the West Pacific Rim. It is hardly surprising, then, that in so much of the literature there is the suggestion, or at least the implication, that many of the West Pacific Rim countries provide a model – a particular set of circumstances which can explain that success. How valid is this notion?

A second group of questions refers to whether the experience of the West Pacific Rim helps to throw light on new approaches or attitudes to the study of economic development. Does their experience support the contention that a clear distinction needs to be made between the cultural – growth with equity and the wealth creation – growth approaches to economic development? For the most part, the countries of the region have in practice rejected the former and embraced the latter. But why have two or three of the countries in our area been less successful than the others? Why are some countries, notably Japan and the four NICs or 'Tigers' of South Korea, Taiwan, Hong Kong and Singapore, so much more successful economically than the other countries? Why has South Korea achieved what has been called 'an economic miracle' while North Korea has remained in the economic doldrums? Has relative success or failure anything to do with the persistence of centrally controlled command economies in the communist countries, whereas other countries have embraced capitalism, with or without associated democratic institutions? Why has the adoption of capitalism and market-oriented strategies meant more equity, less poverty, and better health and education for the poorest sections of the populations in the economically most successful countries? And what does this mean for the dependency, neo-dependency or other similar theoretical analyses formerly so widely prescribed for and applied to other parts of the developing world?

Finally, what geopolitical and geostrategic role does the West Pacific Rim play in the wider Pacific world, including the East Pacific Rim – Canada, the United States and Latin America – and Australasia to the south? If, as suggested in these pages, the countries around the Pacific basin eventually begin to dominate the world economy, what are the broader implications of this trend for Europe and for the global economic system? What, in particular, will happen if China, with its vast resources, population and potential domestic market, ever becomes the latest, and greatest, of the East Asian 'Tigers'? And what implications will the greatly increased economic power of the Pacific Rim as a whole have for global security and the future balance of world power?

The underlying argument and plan of the book

In considering economic growth in the West Pacific Rim, and in particular the three groups of questions referred to in the previous section, it seems necessary to have a clear point of view, without which the discussion might well be obscured by a mass of information, conflicting ideas and unstated prejudices. The argument in these pages can be briefly stated here, pending a more thorough and critical re-examination in the final chapter. It suggests that each economy and each case of economic growth in the West Pacific Rim is in all important respects unique. The corollary of this is to argue that there can be no development model beyond that identified in any particular country; nor can one country's experience be transferred to or replicated elsewhere.

The chapters that follow develop this underlying argument in the following manner. The first chapter begins by establishing the extent to which the various economies of the West Pacific Rim have experienced successful economic growth. The rest of the chapter and the two that follow it examine critically the view, expressed by many writers, that natural and human resources, the historical legacy (especially colonialism), and culture can be said to 'explain' success or failure. It is concluded that none of these factors has any significant explanatory power.

Chapter 4 discusses the case of the Overseas Chinese – the Chinese diaspora – in the West Pacific Rim. These Chinese are considered to be of central importance to the argument because they have provided a major catalyst for economic growth in the various host countries, particularly through their entrepreneurial, commercial and trading activities. This introduces the theme of trade, including both domestic trade (Chapter 5) and international trade and investment (Chapter 6). It is argued that trade has been and remains the *primum mobile* or main generator of economic growth in the most successful countries of the West Pacific Rim.

Chapter 7 brings together the material and ideas from the six preceding chapters by focusing on the economic policies or strategies followed in the various countries; it pays particular attention to the extent and nature of government intervention in both communist and non-communist economies.

The three chapters that follow (Chapters 8, 9 and 10) move away from the rather generalised and abstract discussion in the earlier part of the book and focus, in turn, on Japan and the four Newly Industrialised Countries (NICs, 'Tigers' or 'Dragons') of Singapore, Hong Kong, Taiwan and South Korea; China; and the Southeast Asian sector (Vietnam, Malaysia, Brunei, the Philippines and Indonesia). These chapters are designed to provide rather more detailed country-based material than was appropriate in the earlier chapters. But they also help to emphasise how very distinctive each country is in many important ways, including the manner in which it has chosen to meet the challenge of economic growth.

Chapter 11 places the economic success of the region in its broader Pacific and global context. With the collapse of communism and political disintegration in the former Soviet Union, attention is now increasingly being directed towards the implications of East Asia's economic growth for the world balance of economic power, especially *vis-à-vis* the United States and Europe.

The concluding chapter attempts to answer the three groups of questions posed in this Introduction and goes on to discuss the material presented in these pages against a number of ideas about economic growth and the role of trade in that growth.

Notes

1 For interesting discussions on defining the Pacific Rim, see Gibson, 1990, pp. 1–8; Winchester, 1991, pp. 36–40.

1

Patterns and resources

Patterns of development

Underlying much of the discussion in these pages is the assumption that the countries of the West Pacific Rim have achieved high levels of success in their recent economic development. The conventional wisdom about these countries can be summed up in three statements: (i) they have achieved high absolute levels of economic development; (ii) these levels and rates of growth are significantly higher than in most other parts of the developing world, which means that the real income gap between the developed industrial countries, including Japan, and most other countries in East Asia has narrowed dramatically; and (iii) income inequalities are low, particularly in those countries with the highest per capita incomes. It is clearly important to begin by examining these three statements in order to discover how far they are true, and whether they apply equally to all eleven countries or only to a few.

There are predictable difficulties in finding and using comparative data of the kind required for such a task. Not only are there great variations in the dates and reliability of statistical material, but some countries do not publish data regularly or even at all. Thus in official statistical series, such as those of the United Nations or World Bank, there are serious gaps in the information for China, Vietnam, Taiwan and Brunei. In some cases it has been possible to fill these gaps from local sources, but the problem of reliable comparative data sources remains. There are also difficulties arising from the different criteria used by some governments. In China, for instance, GDP figures refer only to the gross value of industrial and agricultural output, ignoring services altogether.

The statement that the countries of the West Pacific Rim have achieved high absolute levels of development is clearly more true of some countries than of others. Table 1.1 summarises the differential pattern of economic development on the basis of per capita income figures. Using World Bank categories, the general pattern of economic achievement reveals very considerable variations. China, Indonesia and Vietnam are classified as 'low-income economies'; the Philippines and Malaysia are termed 'lower-middle-income economies'; South Korea is an 'upper-middle-income economy'; and Taiwan, Singapore, Brunei, Hong Kong and Japan are categorised as 'high-income economies'.

Table 1.1 Population and GNP (per capita) and growth

	Population (millions) mid- 1989	GNP per capita US$ 1989	Average annual growth rate GNP per capita 1965–89
Low-income economies			
China	1,113.9	350	5.7
Indonesia	178.2	500	4.4
Vietnam	64.8	150*	–
Lower-middle-income economies			
Philippines	60.0	710	1.6
Malaysia	17.4	2,160	4.0
Upper-middle-income economies			
South Korea	42.4	4,400	7.0
High-income economies			
Taiwan	19.8	6,033	7.03
Hong Kong	5.7	10,350	6.3
Singapore	2.7	10,450	7.0
Brunei	0.25	15,390[†]	2.2[‡]
Japan	123.1	23,810	4.3

Sources: World Bank (1991); World of Information (1990).
*   estimated 1986
[†] 1987
[‡] 1988

Thus only six out of the eleven economies are classified as upper-middle-income or high-income economies. In these six economies of Japan, Brunei and the four so-called 'Tigers' of South Korea, Singapore, Hong Kong and Taiwan, levels of per capita income are over US$4,400 a year. However, taken together, these six countries account for less than 10 per cent of the total population in the West Pacific Rim, whereas the three poorest countries of China, Vietnam and Indonesia account for almost 90 per cent of the area's population. For most people in the West Pacific Rim, then, absolute levels of income per capita are by no means high. More precisely, there is a hierarchy of development levels, with

Japan – probably the world's most successful developed industrial economy – well out in front with the four 'Tigers' and Brunei behind, followed some way behind by the other countries. This point is sometimes expressed in terms of the 'flying goose formation'.

The pattern remains virtually unchanged when criteria other than per capita income are used (Tables 1.3, 1.4, and 1.5). Population growth rates, age structures, life expectancy, and several other health and educational indices do little to change the general pattern suggested by per capita income figures. As for 'quality of life', most indicators suggest that this has improved, not deteriorated, with development, especially in the NICs, though less in Indonesia, where life expectancy remains relatively low (Riedel, 1988, pp. 21–2).

At first sight, the low position of the five poorest economies in the World Bank league table suggests that only the six most successful West Pacific Rim economies have done significantly better than other countries in the world. After all, China lies well down below Benin, Zambia and Mauritania, for example. However, if per capita income *growth* rates are taken into account, a rather different picture emerges (see Table 1.1). China and Indonesia have exceptionally high growth rates compared with other low-income economies; and the same is true of Malaysia in the lower-middle-income economies. At the regional scale this point is even more striking. Whereas the average annual growth rate per capita for East Asia was 5.2 per cent for the period 1965–89, the figure was only 0.3 per cent for sub-Saharan Africa, 1.9 per cent for Latin America and the Caribbean, and 1.8 per cent for South Asia.

It is widely accepted in the literature that one of the remarkable features of economic development in the West Pacific Rim is that success there seems to have been achieved without incurring widening gaps in income levels (see Figure 1.1). Such evidence as exists, indeed, suggests that income differentials are being reduced as per capita income levels rise. In Taiwan, for example, economic growth created a shortage of labour, causing low salaries to rise faster than high salaries, so reducing income disparities. The income disparity ratio, comparing the top 20 per cent with the bottom 20 per cent, fell from 15 in 1952 to 4.9 in 1982; today it is 4.2, or lower than in Japan and the United States. This economic growth with equity has played a major role in Taiwan's social change, encouraging mass consumption of such items as telephones and washing machines, improving diet, enlarging the middle class, and improving educational standards (Copper, 1990, p. 46). In Japan, more sophisticated data on this matter are now available. Minami (1986) shows that in Japan, whereas in the pre-war period economic growth led to greater income inequalities, income distribution has become much more equal since 1945.

In this sense the recent experience of the countries of the West Pacific

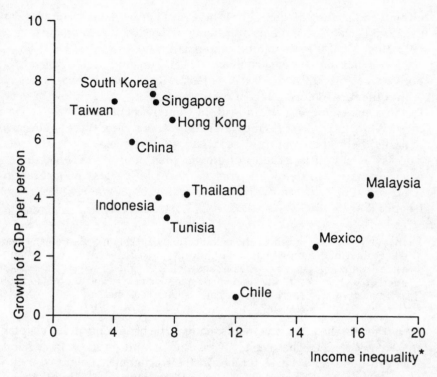

* Multiple of the income of the richest 20% to the income of the poorest 20%

Figure 1.1 Income equality and growth of GDP per person 1965–89 (Source: World Bank; *The Economist*, 16 November, 1991)

Rim in income distribution supports Kuznets (1955) in his view that during the long process of modern economic growth income distribution may at first become more unequal, but that later on the trend is towards greater equality. Riedel (1988, p. 18) notes that East Asian countries have been 'quietly working away to discredit most of the conventional wisdom about the relation between growth and income distribution'. He also demonstrates that many indicators of 'quality of life', such as longevity and school enrolment, have improved most in the countries with the highest growth levels. Dependency theorists, who have sometimes argued that 'open' Third World economies must lead to greater income inequalities, attempt to explain away the case of East Asia by pointing to exceptional circumstances, at least for the four 'Tigers' (Barrett and Chin, 1987, p. 31). However, it seems that the East Asian countries have, by their achievements, worked to discredit the dependency theorists and their views on the relationship between growth and income distribution. Where rapid growth has taken place in the West Pacific Rim it seems eventually to have been accompanied by a fall rather than by a rise in income inequality.

Clearly, it is only too easy to make wild and unsupportable generalisations about the countries of the West Pacific Rim. There is great diversity within them. Absolute levels of per capita income are still low or very low in some countries and, indeed, in some parts of the richest economies. While most West Pacific Rim countries are more successful economically than other countries, especially in the developing world, this 'success' is more true of rates of growth than of absolute levels of per capita income. Furthermore, what is meant by success is a question which must be asked in the East Asian context. As Wade (1989, p. 69) has put it:

> anyone who has experienced the pollution and congestion of East Asian cities will realise that 'success' has to be used in a qualified sense, even leaving aside political aspects such as civil and political rights. I mean 'success' to refer to such basics as the food in people's stomachs and the amount of income left over after caloric requirements have been met.

Nevertheless, the three statements made at the beginning of this chapter are broadly valid. The West Pacific Rim countries have in general achieved quite remarkable success in their economic growth over the past few decades. Compared with other parts of the world – in particular the developing countries in Latin America, Africa and, to some extent, in South Asia – East Asia has experienced rapid economic growth. Throughout the region, moreover, income inequalities are reducing rather than increasing.

In emphasising the speed and extent of recent development in the West Pacific Rim countries, one report (*The Economist*, 1991b, p. 5) has vividly described conditions only forty years ago:

> Asia seemed condemned to poverty. Half of Japan's workforce was in the fields; the other half in factories that were only 15% as productive as America's. 'Made in Japan' meant gimcrack. Shanghai, previously one of Asia's dynamic cities, had fallen to Mao's Communists; its refugees streamed on to a hopeless little island called Taiwan. Korea was poorer than Sudan and on the brink of a devastating civil war.

This is in stark contrast to the picture today, continues the report, in which:

> Japan has emerged as the world's industrial superpower. The four economic 'Tigers' of East Asia – South Korea, Taiwan, Singapore and Hong Kong – have forged the fastest industrial revolutions the world has ever seen. Behind

them are another four countries which are getting close to the point of industrial take-off – Thailand, Malaysia, Indonesia and, most interestingly, China.

Natural resources

The enormous latitudinal extent of the West Pacific Rim, stretching for over 7,000 kilometres from Japan's northernmost island of Hokkaido down to the islands of eastern Indonesia in the south, encompasses a wide range of environments. The southern, Southeast Asian, sector lies wholly within the humid tropics and, for the most part, is significantly wetter and warmer, with less distinct seasons, than in the northern sector; soils and vegetation also reflect this north–south contrast. All these environmental variations are clearly of importance in any examination of the economic development prospects in each country, for each possesses its own unique environmental context – its own opportunities and constraints – within which it has to operate.

With the exception of the interior of China, however, all the countries of the West Pacific Rim possess one crucial environmental advantage over most parts of the developing world, especially in Africa and Latin America. This advantage derives from the region's physical fragmentation into numerous islands, peninsulas and littoral strips. The influence of the sea has been ubiquitous and profound, not only on climate and vegetation, but also on the area's historical development. More will be said on this point in the next chapter, but it is important to recognise that rivers have also played an important part in facilitating contacts between the various physical units and between the area and the outside world. The most obvious cases are the great rivers of Vietnam (Red and Mekong) and of China (Pearl, Yangzte and Huang Ho). But even in the peninsulas and islands, rivers have often been vital in facilitating human settlement and development around their mouths, enabling contact to be made with peoples in the interior.

In this chapter, however, interest is focused on the resource endowment aspect of the differing physical environments. In particular, there is a need to examine briefly the now commonly stated assertion that the natural resource base for development in the West Pacific Rim is poor or negligible in those countries which have been most successful in achieving economic progress; and, conversely, that the countries with the most substantial natural resources have performed least well. This is clearly a matter of some theoretical as well as practical interest.

The country with easily the largest endowment of natural resources is China, though, as Edmonds (1991, p. 1) points out, this does not mean that its per capita possession of natural resources is high. Dealing first with the land resources, of the total surface area of 9.6 million square kilometres in China, mountains cover nearly 35 per cent, plateau land over 25 per cent, hill country 10 per cent, and only 31 per cent is classified as basin and plain. Even more striking, only 10 per cent of the total area is cultivated. The reasons for this, as well as the regional variations in development, will be discussed in more detail in Chapter 6. But it is clear that size by itself is not necessary for successful economic development. Indeed, the most prosperous countries in the West Pacific Rim – Japan, South Korea, Taiwan, Hong Kong and Singapore – are countries with quite small, in some cases very small, land areas; and in Japan the problem of limited level land is perhaps even greater than in China. These successful countries are also all insular or peninsular countries. On the other hand, the two large and most fragmented island countries of Indonesia and the Philippines are relatively poor in per capita terms.

Another factor believed by some writers to be important, even critical, to development is climate. While the cruder forms of climatic determinism may now be discounted, some authorities point to the existence of significant seasonal temperature contrasts in the northern sector of the West Pacific Rim, compared with the relative lack of such seasonal temperature contrasts in the southern, Southeast Asian, sector. Either explicitly or implicitly, it is sometimes suggested that someone living in Indonesia, for example, is unlikely to possess the same level of physical, or even mental, energy as someone living in Japan because of this factor of seasonal temperature change. The difficulty with such a view is not only that there is no reliable scientific evidence to support it, but also that it is obviously not the case in other examples: Singapore – one of the most successful economies in the West Pacific Rim – is located in the southern sector and has a uniformly hot and humid climate probably as uncomfortable as anywhere in the region. Conversely, most of China, near or at the bottom of the development table, is located in the northern sector. However, climate clearly does have a significant effect in other ways: in the damage caused by typhoons (hurricanes, cyclones or *baguios*) and the floods so often associated with them in many countries of the region. This is also a region of great physical instability, with constant threat from earthquakes and volcanic eruptions.

The main resource endowment with which we need to be concerned here, however, is mineral resources. There is a clear inverse relationship between the possession of mineral resources for industry and the levels of per capita income. China has by far the largest and most varied endowment of mineral resources, including coal, oil, natural gas and iron ore. Yet it is among the poorest of all countries in the region. Conversely, wealthy Japan and the four NICs of South Korea, Taiwan, Hong Kong and Singapore have few or neglible mineral resources.

The one exception to this scenario is Brunei, which has both a very high per capita income and is an important oil and natural gas producer. The case of Brunei, however, illustrates very clearly the weakness of using per capita income as a true measure of development, as it does not have the economic structure of a developed, urban–industrial economy (Table 1.2).

Table 1.2 Structure of production, 1965 and 1989 (% distribution of GDP)

	Agriculture*		Industry†		(Manufacturing)		Services	
	1965	1989	1965	1989	(1965	1989)	1965	1989
China	44	32	39	48	(31	34)	17	20
Indonesia	56	23	13	37	(8	17)	31	39
Philippines	26	24	28	33	(20	22)	46	43
Malaysia	28	29	25	21	(9	7)	47	58
South Korea	38	10	25	44	(18	26)	37	46
Taiwan	–	6	–	51	(–	32)	–	43
Hong Kong	2	0	40	28	(24	21)	58	72
Singapore	3	0	24	37	(15	26)	74	63
Japan	10	3	44	41	(34	30)	46	56)

Source: World Bank (1991); World of Information (1990).
* includes forestry, hunting and fishing as well as agriculture
† equals value added in mining, manufacturing (shown as a separate sub-group in brackets), construction, electricity, water and gas

The view that natural resources can and do determine the extent of economic development in a country has sometimes found favour among geographers: indeed, this view is sometimes rather disparagingly referred as the 'geographical determinist' school of thought. It is based on some evidence, of course. Without oil, for example, some Persian Gulf countries could never have achieved the often very high levels of development and per capita income they have achieved. And within the West Pacific Rim, Malaysia and Indonesia, as well as Brunei, have also benefited immeasurably from their oil production capacity.

On the other hand, it is difficult to find in our area any country – apart from the case of Brunei – where the possession of natural resources can in any sense be said to have *determined* levels of development. Indeed, as noted earlier, the most prosperous countries seem to be unusually deficient in natural resources of all kinds. Singapore and Hong Kong, in particular, have no natural resources at all, unless one includes strategic position and, in the case of Hong Kong, a fine natural harbour. Almost everything that supplies and sustains the people and industries of these two territories, including even water and most food, has to be brought in from outside; and Japan, perhaps the most successful economy in the world, is the largest net importer of food among the industrialised countries.

The evidence from the West Pacific Rim countries, then, tends to support the view that lack of natural resources need be no handicap to a country in its pursuit of rapid economic development. As Bauer and Yamey (1957, p. 57) put it many years ago,

> since the changing fortunes of many countries and regions have not been connected with the discovery or exhaustion of natural resources within their territories, the fortuitous distribution of these resources certainly does not provide the only, and probably not even the principal, explanation of differences in development and prosperity.

For after all, natural resources can be substituted by several means, all of which are well documented and are particularly well illustrated in the countries of the region. First, there is trade, perhaps the leitmotiv of economic prosperity in the region. If a particular resource is needed, it can be acquired through trade. This also has the great advantage that it ties in economic growth directly, not to a fixed or diminishing natural resource – as Brunei is already discovering – but to people and capital, both of which have great dynamic potential. Secondly, any lack of natural resources can also be compensated for by substituting capital, technology and labour skills; this includes education and training – 'investment in human capital', managerial capacity and economies of scale – all of which have certainly been critical in the economic development of Japan and the four NICs.

The approach taken in these pages, then, is that in only a very few limited cases is the presence or absence of particular natural resources a significant consideration. In the West Pacific Rim there have been few physical constraints on development, though this is not to suggest that a consideration of the physical environment – the entire natural resource base – is not important for any true understanding of the unique context within which every country's development efforts must operate. Furthermore, it can even be argued that poverty in natural resources can act as a stimulus to the development of other resources. As Minami (1986) suggests, for instance, the lack of natural resources meant that Japan simply *had* to develop its human resources more rapidly and effectively.

Human resources

The basic facts about the human resources are summarised in Tables 1.3, 1.4 and 1.5. Here, again, the case of China stands out very sharply. If the sheer size of a population has anything to do with economic success, then China would be a paradise, containing within its borders as it does between 20 and 25 per cent of the world's population. And Indonesia, with the world's fourth largest population, would be well up the league table of economic prosperity. In fact, however, China and Indonesia are well down the table, whereas countries with small or medium-sized populations are the most prosperous: Singapore has only 2.5 million inhabitants, and Brunei less than half a million. Clearly, then, there is no obvious causal connection – either positive or negative – between population size and development levels.

The same general point can be made about population densities (Table 1.3). Leaving aside for the moment the important qualification of differences within countries, it is clear that the countries with the highest average population densities have the highest income levels. Singapore, a small, densely populated island of over 2.5 million people, and Hong Kong, with over 6 million, have very high per capita incomes. Whether or not high densities reflect or cause development is a topic we cannot address here. But it is worth considering whether it is purely fortuitous that levels of development are highest where population densities are highest. This question impinges not only on the debates about optimum population, overpopulation, underpopulation, and population pressure: it is also relevant to the argument that population pressure may be a necessary precondition for economic change.

As for the distributional pattern of population in the various countries of the area, the same problem arises: how to separate cause from effect. In many cases, topography influences distribution, as in China, Japan, Taiwan, Vietnam and Malaysia. But other cases are rather different. In Indonesia, for instance, the large population is concentrated on the islands of Java, Madura and Sumatra, whereas the Outer Islands are very lightly populated – a fact which, as the next chapter will show, has an historical explanation. The case of Indonesia is also interesting in that it demonstrates the effectiveness or otherwise of attempts to even out the distribution of population in a country. In order to reduce population pressure in Java and to assist in the development of the Outer Islands the Indonesian government has been following a transmigration policy. More will be said about this and other migration issues in later chapters.

One further aspect of population distribution is the distribution between rural and urban areas (see Table 1.3). Here the figures are unusually suspect, especially in any comparative analysis. In the case of China, for instance, the most recent urbanisation figure – the percentage of the total population said to live in urban areas – is 50 per cent. However, this only makes sense if set against the definition of 'urban' in China which includes large areas which are rural but which have been

Table 1.3 Selected demographic data, 1989

	Area sq kms (000)	Population (millions)	Population density per sq km	Average annual population growth*	Crude birth rate per 1000	Crude death rate per 1000	Population % under 15	Urban %	Life expectancy at birth
China	9,561	1,113.9	116.5	1.4	22	7	27.2	53	70
Indonesia	1,905	178.2	93.5	1.6	27	9	36.8	30	61
Vietnam	330	64.8	196.4	2.2	32	7	40.1	22	66
Philippines	300	60.0	200.0	1.8	30	7	40.1	42	64
Malaysia	330	17.4	52.7	2.2	30	5	37.8	42	70
South Korea	99	42.4	428.3	0.9	16	6	26.4	71	70
Taiwan	36	19.8	550.0	1.0	17	5	30.0	67	75
Hong Kong	1	5.7	5,700.0	0.9	14	5	21.6	94	78
Singapore	1	2.7	2,700.0	1.0	18	5	23.7	100	74
Brunei	6	0.25	41.7	3.2	30	4	38.0	64	75
Japan	378	123.1	325.7	0.4	11	7	19.0	77	79

Sources: World Bank (1991); World of Information (1990).

* estimated 1989–2000

placed under the jurisdiction of municipal or urban authorities. The same point explains the 100 per cent urbanisation figure for the city-state of Singapore. Nevertheless, it is clear that the highest levels of urbanisation (over 69 per cent) occur in the most developed and prosperous countries.

Turning to the dynamics of population numbers, inward and outward migration are dealt with in later chapters, but they nowhere significantly affect rates of population growth in the West Pacific Rim. The crude birth and death rates are summarised in Table 1.3. They are particularly interesting because they reveal how far along the demographic transition or cycle the various countries have moved. Compared with most developing countries, crude death rates are low (between 5 and 9 per thousand), reflecting the extent to which disease and health have been controlled compared with the situation in, for instance, sub-Saharan Africa. It is therefore on birth rates that attention must be focused in order to understand the differential rates of population growth and the rationale for the various policies adopted to control population growth.

The countries with the lowest rates of population growth (1 per cent or less: Japan, Hong Kong, Singapore and South Korea) also have the lowest crude death rates (11 to 16 per thousand). And it is these same countries that are the most prosperous, supporting the assumption that high rates of population growth increase the child dependency burden, increase the danger that rates of population growth will outstrip rates of economic growth, and have negative effects on social and economic development generally. This clearly affects the argument over whether population control is a necessary precondition for economic development; or whether it is development itself which leads to effective birth control. Throughout the region, population control measures are widely advocated, though with varying degrees of success. In China, the 'one-child-per-family' policy has been effectively imposed – at least in the urban areas – whereas in the Philippines the voluntary use of contraceptive methods has been less successful, having run into serious opposition from the Catholic Church.

One of the demographic implications of severely reduced rates of natural increase in many countries of the West Pacific Rim is the effect on the age structure. As Japan is now discovering, an ageing population is becoming a characteristic of countries with low growth rates of population, leading to increasing levels of elderly dependency, a problem which must eventually be faced by all countries in the region. In the meantime, however, life expectancy is rising, and as the number of young adults as a proportion of the total population increases, so the number of households increases. With improved standards of living and economic expectations, this must lead to rapid increases in demand for services, including housing. In South Korea, for instance, there were 8.1

million households in 1980; but by the year 2005 the figure is expected to rise to 14.7 million households, most of them in the cities.

In trying to determine how far the numbers, distribution or dynamics of population affect economic growth, it is once again a problem of cause and effect. Perhaps the main comment to make is that the evidence is conflicting and subject to many different interpretations. It seems that only in the case of high population growth rates is it possible to suggest with any confidence that economic development is being constrained. On the other hand, population growth rates throughout the West Pacific Rim are not at all high by world standards, and rarely do these growth rates exceed the rates of economic growth.

Of far more importance in any attempt to explain differential levels of development between the countries of the West Pacific Rim is the quality of the population. Indeed, it may well be that in trying to understand the broader question – why the region has become the centre of gravity of the world economy – an important part of the answer lies, not in the area's endowment of natural resources, nor in its huge populations, but rather in the energies, skills and organisation of its peoples. The cultural determinants in this matter will be discussed at length in Chapter 3. For the moment it is necessary to look briefly at some of the basic indicators of 'quality', especially health and education.

Health is represented in Table 1.4 by several indices: population per physician, population per nursing persons, percentage of births attended by health staff, percentage of babies with low birth weights, infant mortality rates and daily calorie supply. Of the available figures – and there are some serious gaps – Indonesia and the Philippines come off worst, while in most respects Singapore, Hong Kong and Japan do well. As for infant mortality rates, these are remarkably low for Japan, Hong Kong and Singapore. There are rather lower figures for daily calorie supply per capita in Vietnam, Indonesia and the Philippines than elsewhere, but these figures do not reveal anything significant in the absence of figures for the quality of diet. By standards of world comparison, none of these indicators is particularly disturbing, though a good deal remains to be achieved in certain countries. It is also worth nothing the surprisingly good figures for China.

But what relevance has all this for development prospects in the countries of the West Pacific Rim? To some extent, certainly, better health and nutrition may result directly from development. Nevertheless, and quite apart from all moral and humanitarian arguments, the literature does suggest a positive effect of good health and nutrition on productivity, especially in agriculture. In Indonesia, for example, it has been shown that the productivity of workers who received iron supplements for two months rose by between 15 and 25 per cent. A significant link has been demonstrated between wages and

Table 1.4 Selected health data

	Population per physician (1984)	Population per nursing person (1984)	Births attended by health staff (% 1985)	Babies with low birth weight (% 1985)	Infant mortality rate per 1000 live births (1989)	Daily calorie supply per capita (1988)
China	1,010	1,610	–	6	30	2,632
Indonesia	9,460	1,260	43	14	64	2,670
Vietnam	950	590	–	18	43	2,233
Philippines	6,570	2,680	–	18	42	2,255
Malaysia	1,930	1,010	82	9	22	2,686
South Korea	1,160	580	65	9	23	2,878
Taiwan	1,100	–	–	–	9	–
Hong Kong	1,070	240	–	4	7	2,899
Singapore	1,310	–	100	7	8	2,892
Brunei	2,000	–	–	–	12	–
Japan	660	180	100	5	4	2,848

Source: World Bank (1991).

weight-for-height among casual agricultural labourers in Indonesia. Some studies have also shown the positive effects of better nutrition on a child's capacity to learn. Studies in China over many years have consistently shown that protein-energy malnutrition is related to lower cognitive test scores and poor school performance. And in Indonesia a study found that iodine deficiency reduced cognitive performance among nine- to twelve-year-old children.

Education is represented in Table 1.5 by the percentage of age groups enrolled in primary, secondary and tertiary education. The differences between countries is apparent in the secondary sector and very marked in the tertiary sector. These are predictable differences and demonstrate how important higher education is to a rapidly industrialising country. But recent research in the area also points to a very strong link between primary school education and economic growth. Studies in Malaysia and Korea confirm that schooling substantially raises farmers' productivity. Investment in education is clearly worthwhile for many reasons, including the fact that educated farmers are more likely to adopt new technologies and to produce higher returns on their land. And it is not just in the labour market that improvements seem to result from education. It is estimated that one year's education for women can lead to a 9 per cent decrease in under-five mortality. Other things being equal, the children of better-educated mothers tend to be healthier.

On the other hand, 'education' is a broad, rather vague term covering many types and qualities of training. It may be questioned whether

Table 1.5 Selected education data, 1988

	Primary		Secondary		Tertiary	Adult illiteracy 1985
	Total	Female	Total	Female	Total	%
China	134	126	44	37	2	31
Indonesia	119	117	48	43	1	26
Vietnam	102	99	42	40	–	–
Philippines	110	111	71	71	38	14
Malaysia	102	102	57	57	7	–
South Korea	104	104	87	84	36	–
Hong Kong	106	105	74	76	–	12
Singapore	111	110	69	70	–	14
Japan	102	101	95	96	35	5

Note: Figures refer to the percentage of age groups enrolled in education. Gross enrollment in primary education commonly exceeds 100% because some pupils are younger or older than a country's standard primary school age.

Source: World Bank (1991).

education in its narrowest sense is necessarily beneficial in the development process, however desirable it might be to the individual in a cultural sense. After all, is some of the education practised in China or Vietnam necessary to the economic growth of these countries? Education for politicisation, to instil an ideology, to encourage instant obedience and to discourage individualism and criticism of the authorities – is this either desirable or necessary? And if we are talking only of academic education in the Western sense, then does this not encourage élitism and a tendency to despise commercial, agricultural and industrial occupations? A good deal of literature focuses on these questions and on the precise form education should take in any particular country, and it seems that few worthwhile generalisations are possible. It has been pointed out that the role of education and the forms it takes will be very different for a dispersed, rural, Muslim population in Indonesia, for a largely Chinese population in the prosperous city-state of Singapore, and for the peasantry of the Communist-controlled society of China. It is important that education should not create disaffected intellectuals, but it is also important that it should be in close touch with the life of the community as a whole. Education should be in harmony with the technical and administrative requirements of each particular country. This means, for example, that short-term technical training without the luxury of a lengthy academic or more formal education may, in certain circumstances, be the most appropriate form of education.

Nevertheless, the case of Japan tends to provide convincing proof of the critical importance of school education. At the beginning of the Meiji Restoration in 1868 only about 15 per cent of the population of Japan was literate. In 1872, however, elementary school education was made compulsory throughout Japan, and later, in her colonies of Korea and Taiwan, education and literacy were made central planks of development planning.

Other qualities are often written about, but there is very little hard information to go on. This is true, for instance, of attitudes and aptitudes regarding work. Some writers refer to the work ethic (or its absence) among particular groups of people in the countries of the West Pacific Rim. Apart from the climatic determinism argument already referred to, some writers suggest that racial or ethnic differences can explain why, for instance, the Chinese in Malaysia seem so much more thrusting and successful than the indigenous Malays. Is this difference innate? Or is it the expression of some cultural trait, such as religion or sets of values? Chapter 3 will examine this cultural argument in some detail. But at this stage it might be useful at least to suggest that entrepreneurial ability and energy, initiative, independence and an enthusiastic work ethic might reflect not the race, religion or culture of an individial or group, but rather an institutional framework which allows an individual's energies, initiative and ambitions to develop and operate freely. More precisely, does not the case of the Chinese suggest that it is the social, political and economic institutional framework in Mainland China – state control, collectivism and the whole paraphernalia of Chinese communism – which is responsible for China's heavily bureaucratic and inefficient economic performance. After all, the Chinese operating in Hong Kong, Taiwan and Singapore came from China in the first place.

In considering the natural and human resources of the various countries of the West Pacific Rim, there is clearly much one can say, backed up sometimes by rather dubious and uneven evidence, about the relationship between these resources and economic development. But much of it is more interesting than relevant. With natural resources, or with the quantitative or qualitative aspects of human resources, no kind of direct or simple causal connection can be found. Even when discussing the quality of human resources, can it really be suggested that there are differential capacities that make some groups more effective than others in the development process? Do such differential capacities help to explain different levels of development? Yet it is clearly going too far to suggest that natural and human resources are rarely critical, and may even be irrelevant factors in the economic growth of countries in the West Pacific Rim. After all, each government, each economy, has to operate within its own unique resource environment, its own particular combination of natural and human resources and, as the next chapter will show, its own historical legacy.

2

Historical perspectives and the colonial legacy

Pre-European contacts

The most significant underlying fact about the historical perspective of development in the various countries of the West Pacific Rim is the great length, depth and variety of their external and internal contacts. Whether for reasons of strategic, trading or imperialist interests, most parts of the region have long been subjected to migrations, invasions and contacts of one kind or another, both from within and from outside the Western Rim of the Pacific.

It was inevitable that this should have been so. Even before European contacts began in the sixteenth century, the fragmented, insular, peninsular or littoral nature of almost all countries in the region had acted against isolation and encouraged contacts, especially by sea and upstream from the mouths of the major rivers. From its early history the constituent parts of the West Pacific Rim have found isolation difficult and external contacts irresistible. Very rarely was there any major physical barrier to penetration that so characterised other regions, such as Africa south of the Sahara or South America. Rarely, too, was there any large-scale organised opposition to penetration, trade or conquest from what were for the most part small island or coastal states. Above all, perhaps, outside contacts were also inevitable because the area lay athwart contacts by land and sea between the two great early civilisations of India and China.

Thus when the Europeans first contacted the region, there was already a tradition of contact with peoples, ideas and goods from distant parts of the world as well as from neighbouring territories within East Asia. Several waves of migration and contact from China, Mongolia,

India and the Arab world had long been operating in the region, and all have left their mark in different ways on the peoples and cultures of the West Pacific Rim.

The Chinese, perhaps more than any other people, have exerted a powerful influence on the development of many parts of the West Pacific Rim. Vietnam was under Chinese rule for over a thousand years until the tenth century AD; and Korea had the same experience for over 400 years up to the fourth century AD. Largely by virtue of its geography, Japan managed to fend off invasions from the mainland. But further south, the peoples of insular and peninsular Southeast Asia were formed over the millenia by waves of migrations, predominantly of southern Mongolian origin and possibly also from the Indian subcontinent. The Malay peoples, as they are now known, stretch from the Malay Peninsula in the west through to the Indonesian and Philippine archipelagos and the islands of Polynesia.

India's major influence began to be felt in the Malay world through their indirect contact by sea (Fisher, 1971). By the later centuries BC, Indian seamen had established trade contacts with the Malay peninsula and Indonesia and, by the sixth century AD, India had become a powerful force in many areas of the southern parts of the region, controlling commerce from Malaysia to the Philippines. Indian influence held sway in various parts of Malaysia, Indonesia and the Philippines until the end of the seventh century, when a new force, Islam, appeared on the scene. Arab Muslim traders established contact with Malaysia in the seventh century, but it was only with the gradual decline of the Indian states and the spread of Islam amongst Indians that Islam took hold in the Malay peninsula. By the fifteenth century Islam had spread throughout Indonesia to the Sulu archipelago and to western Mindanao in the Philippines.

The spread of Islam coincided with the southwards spread of Chinese influence. Chinese seamen had been exploring the coastal parts of Southeast Asia for some time and seem to have established trading contacts with Arab traders at various ports throughout Malaysia and Indonesia. But it was not until the twelfth century that most of the important kingdoms of Southeast Asia were drawn into trade relationships with China. Under these relationships or arrangements the Southeast Asian kingdoms paid tribute to the Chinese Emperor in return for the right to conduct trade. Small but permanent settlements of Chinese traders were established in various ports which were regularly visited by traders from the southern and southeastern coastal cities of China. But with the growth of trade came greater assertiveness. In 1407 China landed a large fleet in Manila from where expeditions were sent throughout the Philippines and the Indian archipelagos proclaiming sovereignty over numerous petty states. In 1571 the Chinese invaded

Luzon, but by this time the Spanish had arrived in force and the Chinese were eventually repelled. Nevertheless, Chinese trade and economic power grew: Chinese settled in Manila and other commercial centres in larger numbers and soon became indispensable to the host communities.

This is not the place to go into any detail about the pre-European history of the West Pacific Rim, fascinating and illuminating though it is (Fisher, 1971). In any case, it will be necessary to come back to some of its implications in later chapters. But for the moment it is more important to comment at some length on the colonial period. For it is to the colonial legacy in the region, whether European, American or Japanese, that some writers point in trying to explain both the differential development and, at least until recently, the relative lack of economic development in the countries of the West Pacific Rim, almost all of which have experienced some form of colonial control in the past.

The development of colonial contacts

The development of colonial control in East Asia has been written up in many excellent sources (see, for instance, Fisher, 1971). Here it is necessary to be aware only of the variety and timing of European penetration in the different parts of the region and to assess how far the colonial legacy has affected development in its constituent countries.

The Portuguese were the first Europeans on the scene: they seized Malacca in 1511, half a century before the Spanish arrived in the Philippines in force in 1565. The Dutch East Indian Company established its headquarters in what is now Jakarta in 1619, although it was not until 1798 that the Dutch government took control of the Dutch East Indies, later to become Indonesia.

British interests started in earnest rather later, towards the end of the eighteenth century, when the East India Company took control of certain strategically selected islands: Penang, which was annexed in 1786; Singapore in 1819; Hong Kong in 1841; and Labuan in 1846. For a brief period, moreover, the British annexed the Dutch East Indies. Throughout the rest of the nineteenth century and early twentieth century the British extended their control over the rest of the Malay peninsula and what is now Malaysia.

The French and Americans began to develop their own interests in the area in the latter half of the nineteenth century. The French took control of the south of Vietnam in 1860 and the northern part in the 1880s; and the Americans found themselves in control of the Philippines after the Spanish–American war of 1898. Finally, Japan's colonial or imperialist ambitions were fulfilled further north, in Manchuria, Korea, parts of

China and in Taiwan, and in the Second World War, they overran much of Malaysia, Singapore, Hong Kong, Macao, Indonesia and the Philippines.

While economic motives, trade for raw materials in particular, have always been important in the process of colonisation, there have also been strong political and strategic reasons for the moves, counter-moves and annexations that marked the colonial period in East Asia. But the precise details of how and why colonialism developed in this region are of less interest to this present discussion than the simple fact that all the countries in the area have at some time been colonised or occupied in one way or another. Each country has its own experience of colonialism, invasion or occupation and this forms part of its own history. What is important to discuss here is how far this experience can help to explain present-day opportunities, problems and attitudes.

An interpretation of the colonial experience

It is commonly suggested that the colonial experience resulted in the various indigenous peoples and economies being exploited, 'undeveloped' and made 'dependent' by the colonial powers in the region. This viewpoint has been widely discussed in the literature and has had a significant impact, not only on academic analyses, but also on political practice and development planning in much of the Third World.

Essentially this approach argues that something profound lay behind the machinations and the acquisition of territory in the process of colonial expansion. According to this view, colonialism was the inevitable consequence of monopoly capital: it involved essentially the export of capital in order to expand markets and to satisfy the desire for ever-increasing profits. Colonialism was exploitative and both caused and perpetuated conditions of poverty and underdevelopment in the colonial territories. All those territories which fell under colonial control were subordinated to the capitalist structures of the Western world and so were geared to meet the requirements of the industrialised nations.

An important corollary of this, so it is often argued, was the creation of a tiered hierarchical structure, with the colonial peoples at the top and the indigenous peoples at the bottom. In most cases the upper levels of the economic structures, including foreign trade, financial institutions, local manufacturing and processing, mining and large agricultural enterprises (all of which were allied with large financial and industrial interests in the metropolitan countries) were dominated by the colonial powers.

In most of the colonial territories of East Asia the middle tier in the hierarchy was formed by the Chinese who, in Malaya, were sometimes even able to compete with or to complement the large European firms. The Chinese monopolised rice milling and the rice markets in many countries; they controlled internal trade in Indonesia; they controlled the export of rubber from Malaya; and they controlled the retail trade of the Philippines. Moreover, the Chinese owned nearly all small and medium-sized factories and controlled the import and export of small-holder products, notably rubber. Below the Chinese were the Indians – the money-lenders, barkeepers and storekeepers – though unlike the Chinese, the Indians were often brought into the lower ranks of the Civil Service. Finally, at the bottom were the indigenous people, the labourers, the peasants and smallholders. They, too, were inevitably drawn into the international capitalist system. In Malaya, for example, about two-thirds of the rubber was produced by indigenous small-holders whose operations were low-cost, flexible and relatively more efficient than in the plantations.

This structure varied from country to country. In the Philippines, for instance, the Filipino landholders were used to produce the necessary export goods – tobacco, manila hemp and sugar cane – the export of which was controlled by American and Chinese firms. And in China, Vietnam and Korea, indigenous landholders and capitalists were similarly used to enable the colonial powers to direct and subordinate the territory and its indigenous peoples to the needs of the colonial power.

According to this interpretation of the colonial period in the West Pacific Rim, then, there were four main consequences. First, the colonial territories became dependent on world markets; this made them vulnerable to changes in world market demand and prices for their primary products, changes over which they had no control. Moreover, in most countries exports comprised a very narrow range of products: tin and rubber in Malaya, coconut oil in the Philippines, and rubber in Indonesia. Colonial territories also became highly dependent on their metropolitan countries. Thus in the Philippines, while the Americans supported free trade, the trade and tariff system was arranged in such a way that most exports went to and most imports came from the United States. The production and export of cash crops was encouraged, largely at the expense of food crops, and overseas trade was conducted largely by non-Filipinos. There was also very little encouragement for manufacturing industry. In some countries, indeed, there was a deliberate and conscious refusal to develop any industry which could compete with that of the colonial power; and it was partly because of this that existing agricultural and handicraft industries went into decline.

Secondly, colonialism resulted in a very unequal distribution of wealth, much of which was concentrated in the hands of the expatriates, Chinese, Indians and, in some cases, small elite cliques from the upper strata of the indigenous peoples. These groups tended to concentrate in a few large centres, usually seaports, which came to dominate the political and social, as well as the economic life of the country. Ancient, traditional, inland capitals were often by-passed and allowed to stagnate. Economic progress did not penetrate very far inland. All this resulted in small, commercialised, coastal zones of development, islands in the ocean of predominantly poor, subsistence economies. This was true almost everywhere, but nowhere more starkly than in China, with its contrast between town and countryside, and with wealth concentrated in the hands of foreigners and a small number of Chinese capitalists, landlords, merchants and industrialists.

While economic progress was largely confined to restricted coastal areas in most of the territories under colonial control, the damaging effects of capitalism nevertheless infiltrated into those inland rural areas where export cash cropping was introduced. Smallholders were tied to the vagaries of world markets. To commercialise and progress they had to borrow money, but found it difficult to do this; they ran up debts; and when they did make profits this was often siphoned off in debt repayments or rent.

Thirdly, it is commonly argued that the colonial powers distorted and aggravated pre-existing regional differences within their territories. In Indonesia, for example, the Dutch first concentrated their attention and resources on Java, which consequently developed metropolitan characteristics and came to dominate the Indonesian archipelago. As far as the rest of the territory was concerned, the Dutch only really wanted to maintain small scattered footholds, outposts to keep other Europeans out. Java was run like a business concern, and the Javanese were compelled to grow and export selected crops, especially sugar and coffee.

Another example was Vietnam, where the French greatly aggravated differences between north and south. While both parts shared a common basic culture, the north, with a climate similar to that in southern China, had acquired a rice-growing economy and a high population density. The southern part, centred on the Mekong delta, was originally a pioneer settlement area settled by peasants coming down from the north. In 1803 North and South had been united under one kingdom, but in the 1860s the French took control of the South, developed the delta and made Saigon the new capital port. They encouraged commercial rice cultivation and made great efforts to modernise its production systems. Immigrants from the south-central part of Vietnam moved down to the south, and Chinese immigrants also moved in and took on the same entrepreneurial and middleman role they enjoyed elsewhere in Southeast Asia. The French established an influential class of landowners who rented land to the peasants.

Even when they annexed North Vietnam, the French continued to concentrate on the development of the South and managed to broaden its export base. But the North, with its high density of population, was unable to increase significantly its export crop production. The French therefore encouraged the manufacture there of consumer goods, using the available cheap labour, and these goods were then sold in the South where the food needed by the North was grown. By the end of the first half of the twentieth century, the French had created an impoverished and embittered North, densely populated and dependent on the South, which was dominated by wealthy landowners and Chinese merchants.

A fourth main consequence of the colonial period – to which attention is commonly directed by those who take this firmly anti-colonial line – is that in most cases the colonial powers created arbitrary international political boundaries. On the one hand, artificial, plural societies were created with an uneasy ethnic and tribal mix; while on the other hand natural affinities were broken.

This school of thought admits that the colonial period resulted in some improvements. Roads and railways, bridges and irrigation works, agrarian research institutes, western medicine, western education and law and order were introduced. But, so the argument goes, the benefits such improvements produced among the indigenous populations were purely incidental. The colonial powers were not motivated by altruism, nor did they attempt to deal with any problems these so-called improvements brought. For instance, western medicine, the clearing of swamps and forests, law and order all reduced death rates; and, as the birth rates remained high, so populations increased dramatically. But the colonial powers did little to help control this population explosion, nor did they help to restructure the economies of their territories so that they could cope with the increasing poverty brought about by these large population increases.

It is also admitted that a distinction should be made between the different forms of colonialism and their different legacies. The Japanese colonial policy, for instance, was very different from that of the Europeans: the Japanese did not follow a policy of encouraging trade dependency in either Korea or Taiwan. Thus the present development of Taiwan must be understood in the light of its experience of Japanese colonial rule for the fifty years from 1895 to 1945. The Japanese developed the country's resources in a manner very different from that followed in most other colonies. Basing their colonial policy on their own early experience of development in the 1870s, Japan practised 'revolution from above', which involved from the very beginning a major cadastral survey and land reform (Wade, 1990, p. 73). Wade notes that absentee landlords had their land rights transferred to local landlords who then supported the Japanese. A good communications infrastructure and improvements in irrigation and fertilisers were used to develop the smallholder food crop production of rice and sugar, items required in Japan. This policy was followed firmly, 'sometimes even with the aid of the police force' (Ishikawa, 1967, p. 102). The Japanese colonial policy was to encourage the smallholder cultivation of staple food crops. It also differed in its industrial policy. Instead of taking out raw materials for processing at home, the Japanese took the industries to the raw materials and labour in Taiwan (Cumings, 1984, pp. 12–13). Indeed, manufacturing growth in Taiwan was sometimes higher than in Japan, and welfare on the island improved quite dramatically (Wade, 1990, p. 74).

By comparison, Indonesia's colonial experience was very different, quite apart from the fact that it had easily the longest period of colonial rule in East Asia. The Dutch in Indonesia tried to control from Europe a vast archipelago of great environmental and ethnic contrasts in Southeast Asia. Their decision to establish their local headquarters on Java was logical, in view of its strategic and historical significance. But the Dutch concentrated their attention almost exclusively on Java up to the end of the nineteenth century and used various repressive measures to induce the local population to export crops, especially sugar and coffee. After the turn of the century, however, the policy changed and the Outer Islands became more important, both for export crops and for the production of petroleum and tin. But still the main interest of the Dutch was to regard Indonesia as a source of wealth for its own metropolitan development.

An alternative interpretation

No doubt there is much that is true in the facts presented above. Clearly, too, the above approach to the study of development is attractive because it can be used to explain everything (especially everything undesirable) that took place during the colonial period. It is also useful in that it can be used to explain what is regarded as slow economic progress since the end of the colonial period, and to explain why the economic prospects for a particular country are so poor. It is thus in many ways a seductive way of treating colonialism and its influence on development. There is no shortage of historical facts and interpretations which can be selected to fit neatly into this explanation; and indeed this interpretation of colonialism has been used by many governments and writers to explain the problems these countries face, to advocate certain solutions, mostly of a predictable political flavour, and to cultivate guilt among citizens of the former colonial powers. Certainly the whole approach outlined above is very strong on rhetoric. Evidence to counter much of this kind of anti-colonial argument, at least as it applies to the countries of the West Pacific Rim, will be presented in several of the chapters that follow. But it is worth making a few general points at this stage.

First, the argument that colonial territories were deliberately under-developed and that exploitative neo-colonial dependent relationships were forced on countries after independence, albeit in a more subtle way, simply does not stand up to the empirical evidence in the West Pacific Rim. Like so many of the arguments on colonialism and its iniquities, it seems to be based on a very narrow range of examples, mostly taken from Tropical Africa.

Hong Kong is a case in point. Still a British colony until 1997, the territory has by any criteria achieved remarkable growth, placing it well up in the list of 'high-income countries' as defined by the World Bank. In Table 1.1 the colony comes next to Japan in per capita income, and just above Singapore and South Korea. Singapore broke away from the Malaysian Federation in 1965; and South Korea suffered a devastating war. Yet both countries have prospered astonishingly. Japan, too, is an interesting case in this respect. Although never made a formal colony, Japan at the end of the Second World War was probably in a worse state than any of the colonial or former colonial territories in the region. The country had been militarily and, it seemed, psychologically crushed; and its economy, industry, agriculture and commerce, was in ruins. Moreover, in the first few years after the war, the occupying American forces actively dismantled the remaining heavy industry. Finding enough food and shelter was the primary concern of most Japanese. No one could have predicted that within twenty years or so Japan would become the third largest industrial power in the world and that within fifty years it would stand alongside the United States in economic power and prestige.

As for other countries in the West Pacific Rim, while they may not have achieved such dramatic increases, Malaysia, Indonesia and the Philippines have shown very respectable rates of growth, well above those in Africa. Throughout the region, indeed, it is nowadays difficult to find evidence of the kind of conscious underdevelopment so dear to the hearts of anti-colonialists.

Secondly, the view that the limited achievements of colonialism were largely incidental and were not motivated by altruism or by any real interest in the country seems churlish and unjustified. Guessing at the motives and making judgements about the actions of people who lived in a different world decades, even centuries, ago is a questionable exercise. Even a cursory reading of diaries and accounts written at the time by colonial officers gives a very different picture. It is a field where retrospective moral judgements are all too common. Of course there was much to criticise about colonialism in practice. But there was also much to applaud and to make use of in independence.

Singapore is a good example of a country whose government accepted the fact of its colonial past at independence and determined that it would build on that past and never use it as an excuse for failure. Singapore had the self-confidence in independence to preserve what was good in its colonial past and to reject what it disliked. In Korea, despite the enmity and hatred felt towards the Japanese – who had not only occupied Korea but had also tried systematically to eliminate its culture – there was much that the Japanese left behind which could be and was used constructively. Many Korean laws and administrative procedures date back to the Japanese colonial period; and Japanese modern farming, irrigation, and agricultural research services formed the basis on which Korea was able to expand and modernise its rural sector. Some Japanese industries, too, were taken over by Korean entrepreneurs who had worked in these industries under the Japanese in more junior positions.

Thirdly, the argument that colonial territories have suffered from being tied into the internationalist capitalist system is a criticism which seems again to have little relevance to the countries of the West Pacific Rim. Indeed, as a later chapter will argue, international trade and foreign investment have been major factors – perhaps *the* major factors – in their growth over the past forty to fifty years. Those countries which have deliberately kept themselves isolated from the international economy, China and Vietnam, have made least progress in their economic development. It is also significant that both countries have recently rejected their former policy of isolation. China has followed an 'open-door' policy since the late 1970s, and Vietnam has done the same since 1985. In both cases, opening up to international trade has already begun to pay dividends.

The fact, then, that colonialism brought countries into the world capitalist system has in reality been partly, if not largely, responsible for much of the economic success of many countries in the West Pacific Rim. It is true that over-dependence on world markets for too few export products, mostly primary products, has created difficulties. But diversification and domestic industrialisation is enabling most countries to deal effectively with this problem. As for the suggestion that trade is still controlled by the former metropolitan countries, where is the evidence for this? As we shall see in a later chapter, it is in fact remarkable how little trade is now carried on with the former colonial powers in the West Pacific Rim.

Conclusions

The effects of colonialism upon the West Pacific Rim countries were undeniably profound. Boundaries had been changed and in some cases new countries had been created. Populations were now very mixed and often comprised large immigrant groups with considerable political or economic power. Settlement patterns; networks of roads, railways and communications; international trading links, agricultural systems and many other aspects of economic life had been altered by the colonial powers and provided the basis, whether for good or ill, for future development in the former colonial territories. It was then up to each independent government to take action and to react to existing internal and external stimuli and opportunities. Nothing that was to happen was pre-ordained. Our reading of colonial history, like all the history of East Asia, can help us to understand those circumstances. But it cannot reveal why certain decisions were then taken, or why some countries, governments or individuals have succeeded while others have failed.

It is significant that, with few exceptions, there has never been in the countries of the West Pacific Rim the same intense ideological critique of colonialism as that experienced in Africa and Latin America. Why this should be is an interesting question in itself. Some writers have noted this 'unfortunate failure' of scholarship in the region to take up the 'vigorous and theoretically innovative debates that have been part and parcel of the analysis of change in Latin America . . . and Africa'. This failure 'is largely the consequence . . . of the extraordinary influence of positivist and empiricist traditions upon Southeast Asian studies' (Higgott and Robison, 1985, p. 3). But a simpler, and probably more accurate, explanation is that the countries of East Asia have a long tradition of external contacts, of trade and other opportunities, all of which have made its peoples generally less parochial, more pragmatic, more thrusting, and more likely to look towards the future than to be weighed down with an obsession and resentment about their colonial past. Historical determinism, like geographical determinism, has elsewhere spawned a host of theories, political systems and ideologies, many of which are now rapidly being discredited. The most successful countries in the West Pacific Rim today are those which accepted circumstances at independence as they were rather than as they would have liked them to be. They have subsequently channelled their energies into building successful economies and societies on the realities of the present rather than on the basis of an embittered rhetoric about the past.

3

Culture and development in the West Pacific Rim

The cultural perspective is believed by many writers to be central to any study of the development of the various countries of the West Pacific Rim. This may seem axiomatic, if culture in this context is defined as 'the binding element that ties individuals together through their integrated patterns of behaviour, thought and communication, and as such it acts to include some individuals within the group in question and to exclude others from it' (O'Malley, 1988, p. 328). In this sense there is obviously a difference between the Japanese and Malay 'cultures', and on the face of it, that difference might reasonably be expected to exert at least some influence upon the nature and speed of development in these two very different countries. For East Asia as a whole a number of writers have tried to explain its recent economic success as deriving fundamentally 'from some mental or spiritual advantage that Eastasians have over us' (Hofheinz and Calder, 1982, p. 21).

However, there are strongly divergent views about the role of culture in development. On the one hand there are those, including many economists, who argue that culture plays no positive role at all in the development process, though it may have a negative role in the sense that it hampers, inhibits or prevents economic development from occurring. Moreover, the role of culture in development is an immensely broad and complex topic, without clear disciplinary boundaries, so that no one person can possibly be qualified enough to analyse convincingly both the economics and the culture as well as all the links in between, if indeed there are any such links. Cultural explanations are also dismissed by some writers as reflecting an inability to explain development in any other way: 'for those who are baffled by the plethora of conflicting explanations and contradictory evidence, there is refuge in the cultural factor' (Riedel, 1988, p. 2).

The opposing viewpoint argues that culture is an important, perhaps *the* critical factor in the economic development of the countries of the

West Pacific Rim. As Gourevitch (1989, p. 11) puts it, in such a diverse region,

> the expression of cultures is overwhelming. We find in this region one of the oldest civilisations known, China, and some of the youngest . . . With such richness, the possibility of explaining important phenomena through the causality of culture seems irresistibly tempting . . . Culture has been mobilised to explain many if not all of the phenomena of great interest to us.

While it is admitted that culture is not an easy concept to pin down, it is argued that 'to ignore culture largely on the grounds that it is awkward to deal with is both intellectually unsatisfying and potentially costly' (O'Malley, 1988, p. 327).

In tackling this issue it seems useful to begin by looking at a few specific cases, summarising very briefly – and necessarily rather superficially – how some writers describe the cultural characteristics of the major peoples of the region (the Chinese, Japanese, and Malays) and the significance of these characteristics for economic growth.

The Chinese

In the West Pacific Rim the Chinese are numerically easily the dominant people, not only in China, but also in Hong Kong, Taiwan and Singapore. They also form very significant minorities, though usually fairly discrete communities, in Malaysia, Indonesia, Vietnam, Brunei and the Philippines.

The Chinese, whether in China or overseas in the region have a reputation for being energetic and pragmatic and for possessing an all-pervasive, obsessive and uncompromising work ethic. The individual owes his loyalty first and foremost to the family. The family provides support and applies pressure on the individual to achieve professional and material success and expects, indeed often demands, financial and material support in return. In many respects the relationship may be regarded as an informal but nevertheless very strong and binding contract to which private emotions must take second place. Democracy is sometimes regarded as a rather irrelevant notion, and the desire to put the individual's freedom to express his interests and opinions above all else as a rather selfish attitude. Furthermore, political consciousness of society as a distinct and important entity is thought to be an alien and unnecessary concept. These characteristics, combined with the pragmatic necessity of working together, have tended to encourage a strong, paternalistic government, which reinforces the family and its pressures on the individual to prosper for the family or group.

Despite Maoist philosophy, and leaving aside for the moment the question of Confucianism, these same characteristics – a sense of family, the marginalisation of society, and the pragmatic necessity of working together – continued to hold sway in Communist China after 1949 through a strong central government. But unlike in the island states of Taiwan, Hong Kong and Singapore, the logistical imperatives of holding together, managing and defending a vast continental state like China with over a billion people has meant that a tension inevitably exists between these strategic considerations and a natural tendency towards the growth and domination of commerce; and this tension was as true of pre-Communist China as it is of Communist China today (Solinger, 1984). The drive towards commercial integration and economic modernisation in China since 1978, and the strong retroactive forces encountered, is but a contemporary example of this long-existing tension.

China's political system has been said to reflect the inevitability of strong government in the country and in those other territories where Chinese dominate. In discussing the political system in Taiwan, for instance, it has been noted that among the Chinese

> the cultural definition of leadership emphasised . . . the didactic and initiatory role of the leader, who is assumed to lead because of his superior knowledge. This reflects a characteristic pattern in the history of China, of a leaning toward a central organisation and a central authority figure who sets the legal and moral pattern. For centuries one man, the emperor, was seen as the source of all political authority. (Wade, 1990, pp. 229–30)

The Chinese are also said to have a keen sense of racial identity. This feeling, which transcends nationalism and material ambitions, but takes second place to the family, provides a degree of social cohesion. As a trait it is not confined to the Chinese, nor has it yet expressed itself in serious xenophobia or in aggressive racism at the national level. But it is a trait that helps communities to remain distinct or discrete, independent and self-supporting in the host countries elsewhere in the West Pacific Rim. It also provides a transnational cohesion, or at least a sense of common past and future, which at the national level provides a flexible bond within which the family and the individual can operate. It helps to foster a mixture of individualism, strong family cohesion, and strong paternalistic government which can either allow the people to operate with limited interference, or can achieve absolute control.

The Chinese have had a great influence on the cultures of their neighbours. In Vietnam, the literate culture is largely Chinese, and Chinese political systems and constitutions were either imposed by the occupying Chinese or, after the tenth century, copied by the Vietnamese. In Korea, too, many aspects of China's political institutions, religion, philosophy and language permeate Korean life and culture today. Japan has also unconsciously or consciously absorbed and emulated many aspects of China's culture: technology, architecture, religion, political institutions and philosophy.

The Japanese

The Japanese share many of the cultural characteristics of the Chinese. They are similarly energetic and pragmatic and possess the same pervasive and uncompromising work ethic. The family is also of great importance, exerting its pressure on the individual to achieve professional and material success, providing support and expecting loyalty and support in return. The luxury of sentimentality, egotistical individualism and an obsession with one's own emotions and ambitions takes second place. But the Japanese are unlike the Chinese in that they possess a stronger sense of a tangible society, of a nation striving towards a greater whole. The Japanese early regarded themselves as being very different from the Chinese, a distinct people with a quite separate destiny.

Certainly the Japanese have maintained and developed their own indigenous culture. But they have also adopted many aspects of Chinese culture, sometimes improving or adapting them to such an extent that the fact they were originally Chinese was forgotten or regarded as irrelevant. This ability to adopt, adapt, emulate, improve and absorb others' views, ideas, techniques and institutions has become an important feature of Japanese culture.

Cohesion and co-ordination are today very important attributes in Japanese life. Yet effective national cohesion and government developed only very slowly, for the Japanese are an island people, living in a physical environment which inevitably fostered an emphasis on class, tribalism and social fragmentation. A centralised Chinese-style administrative system was adopted in the seventh century, but over the next five centuries Japan moved towards a feudal system: baronial families supported by their *samurai* and running their own mini-kingdoms. This system provided effective control during the twelfth, thirteenth and part of the fourteenth centuries, but then degenerated into war-lordism until the end of the sixteenth century when a Chinese-style centralised administrative system was superimposed over the feudal structure. This lasted until the Meiji Restoration of 1868.

The natural tendency towards co-operation and cohesion at subnational levels; the experience of centralised national control; the ability to emulate; and the traditional subservience of business to government, provided the government after the Meiji Restoration of 1868 with the means to transfer loyalties to the nation. Here was a nation which had been under threat in the past yet had been miraculously saved by divine intervention and which was again under threat from powerful western nations which insisted on opening up Japan to world trade and contacts. The Japanese, under the Meiji regime, found the energy and ability to direct the nation towards a particular goal, industrialisation, and to adopt and utilise various other aspects of westernisation. Unlike the Chinese, with their emphasis on central leadership based on superior knowledge, the Japanese understanding of a leader is that his function is to 'represent the collectivity to the outside and to coordinate interpersonal relations inside' (Wade, 1990, p. 230).

In spite of its physical fragmentation into islands and often quite narrow coastal plains, Japan's culture is remarkably homogeneous. Apart from the small Ainu minority, the Japanese are all Mongoloid and share the same language; its old social class system was abolished after 1868; and in more recent times, after the end of World War II in 1945, even the great *zaibatsu* were broken up. Whether or not this helps to explain Japan's economic growth, it is understandable why the Japanese are as certain about their cultural superiority as are the Chinese. One needs only to recall the *Bushido* theory, that Japan's successful economic performance today 'results from personal or cultural values lacking in the outside world: powerful loyalties to superiors, unstinting attention to results, and above all valour in the combat of competition' (Hofheinz and Calder, 1982, pp. 21–2).

The Malays

The Malay people were formed over the millenia by waves of migrations, and they were influenced by an enormous range of different cultures and beliefs which have intermingled and blended together. The ancestors of the present-day Malays probably originate from southern China and Indo-China or possibly from the Indian subcontinent. In prehistoric times Negroids, Australoids and then Proto-Malays moved east and south down through the Malay peninsula, through Indonesia, north again through the Philippines and the Polynesian islands and, some believe, down further south to Papua New Guinea and Australia. This trek through the Malay peninsula, Indonesia and then northwards again to the Philippines was later followed by the Deutero-Malays and eventually by the southern Mongoloids (from whom the present-day Malays largely originate), who also moved still farther north to Taiwan and Japan (Fisher, 1971).

The diversity of cultures which sprang from these movements and the contrasting environments of these sprawling archipelagos were made still more complex by a wide range of other cultural influences. Relationships with palaeolithic Europe and China have been suggested, but stronger evidence is that the ancient cultures of the Indian subcontinent provided the earliest influence on Malaysia, Indonesia and the Philippines. After the gradual decline of the Indian states and the spread of Islam amongst the Indians, Islam took hold in the Malay peninsula and by the fifteenth century had spread throughout Indonesia to the Sulu archipelago and western Mindanao in the Philippines. The particular brand of Islam that emerged had already to some extent become Indianised and was in any case a tolerant and adaptable philosophy which had no difficulty in accepting and integrating with traditional indigenous beliefs and customs. It soon became and has remained to this day the dominant religion or philosophy in Malaysia and Indonesia and in parts of the southern Philippines.

The Chinese also began to influence markedly the cultural mix of Southeast Asia by trade and by attempted conquest. Chapter 2 noted how expeditions were sent throughout the Philippines and the East Indian archipelagos to proclaim Chinese sovereignty over numerous petty states. In spite of setbacks, Chinese trade and economic power grew: they settled in Manila and other commercial centres in larger and larger numbers and soon became indispensable to the host peoples.

The history of the Malay people of Malaysia, Indonesia and the Philippines is one of diversity and constant movement and change. They are essentially an island and peninsular people and this has made for cohesion at local levels and has reinforced the strength of the family, though rather less so than in the Confucian cultures to the north.

Furthermore, in the dominantly Muslim and Christian world of the Malay peoples, an individual's performance in life, and place in death, is held to be of more significance than the relationship between superior and subordinate persons (O'Malley, 1988, p. 338). Some writers view such a statement as a crude rationalisation of the Malays' relaxed attitude to work and material achievement. But whatever the truth of the matter, it clearly affects their potential for rapid economic growth and has been incorporated into several other interpretations, such as that of Boeke (1966). His view on the indigenous peoples of Indonesia (Dutch East Indies) was that the indigenous, colonised peoples there possessed cultural values that were largely inimical to economic growth. This, he argued, was one reason why the territory developed what he called 'dualism': one part of the economy involved the energetic, economically successful colonialists, together with a few exceptional individuals; the other part was formed by an indigenous population, the bulk of which was held back from participating effectively in the process of economic growth by their own cultural values and expectations.

Interpretation and evaluation

Although there may be some truth in these cultural explanations, it is difficult to accept them as valid explanations for different levels of economic success. Not only do they seem to reduce individuals, human groups, nations and history to mere cyphers or caricatures, they cannot even explain behaviour or why people think as they do, much less determine behaviour and ways of thinking. Still less can cultural elements be regarded as 'factors' affecting or determining economic performance. This is true even of Confucianism.

Confucianism is considered by many writers to be the key to understanding the economic success of East Asia. It is suggested that much of East Asia has been so successful economically largely because so much of it is Confucianist; in the Philippines, by contrast, Christianity has held back the country's development. Lee Kuan Yew has explained the economic success of the Chinese in Singapore in terms of Confucianism: 'the vital intangibles . . . belief in hard work, thrift, filial piety, national pride' (*The Economist*, 1991b, p. 6). Mirroring Weber's 1958 Protestant-ethic argument for Western Europe at the beginning of its Industrial Revolution, it is argued that Confucianism 'instils values such as hard work, thrift, goal attainment, and family loyalty, among others, that prepare people for effective behaviour in market economies' (Gourevitch, 1989, p. 11). Downton (1986) suggests that 'societies with Confucian-Buddhist roots are proving more effective with the industrial and technological challenge on the eve of the 21st century than . . . are those countries with a . . . predominantly Christian-Hebraic tradition' (quoted in Winchester, 1991, p. 61). Winchester goes even further, suggesting that Confucianism 'is central to the coming triumph of the region' and goes on to note that

> the spreading ethic of Confucianism – exported in the last one hundred years or so to every nation on and within the Pacific coastline by the tens of millions of Overseas Chinese who have acted as its accidental evangelists – is crucial. It can be argued that it has played a significant – one might even say the most significant – role in producing all those patterns and distant harmonies that are being recognised as signalling the onset of the Pacific Age (1991, p. 61).

Even allowing for the hyperbole, it is difficult to see how such statements can be accepted as valid explanations for economic success or failure. They ignore several facts. Whereas Confucianism is today being used to explain economic success among the Chinese, at least outside China, it was earlier confidently argued that Confucianism inhibited modernisation and would keep East Asia far behind the West (Wright, 1966). Only thirty or so years ago it was common to blame Confucianism for East Asia's backwardness; it meant:

deference to authority, resistance to change, respect for the ways of the past, aversion to risk, and other values that inhibit responsiveness to the market . . . If culture has explanatory power, it must be relatively constant. If economic change occurs rapidly, the constant cannot account for the change (Gourevitch, 1989, p. 12).

Again, many countries elsewhere have achieved success without being Confucianist. And China, the heartland of Confucianism, is only now beginning its industrial transformation. Anticipating slightly a later discussion in these pages, it is worth referring to the comment by Vogel on Confucianism: Just as Max Weber found that the

> greatest drive to industrialise in his time came in areas located far from Catholic orthodoxy, so in East Asia industrialisation prospered in areas far from the centres of traditional Confucian orthodoxy, where trade and commerce were most highly developed (Vogel, 1991, p. 84).

Conclusions

It is argued here, then, that cultural interpretations of why people think and behave as they do, and why they organise their economies and political structures as they do, are of little help to us in our attempts to understand why a particular economy has been successful while another has not. Analyses which dig deep into a country's philosophy, religion, customs, magic and the broad sweep of its history in an attempt to explain why people in any part of the West Pacific Rim come to think, behave and organise in the way they do are of course fascinating. The cultural characteristics they reveal may indeed help or hinder development. But they cannot explain or contribute towards an explanation of that development.

It may be that part of the problem in any attempt to explain economic growth in the West Pacific Rim in terms of culture is that the discussion is always on too general a level. Perhaps, as one writer puts it, 'we need to handle only specific cultural elements in relation to specific economic actions and decisions, instead of . . . talking vaguely about culture and development as whole concepts' (O'Malley, 1988, p. 343). On the other hand, Hsiao argues against any kind of specificity:

> cultural factors should not be interpreted as individual social behaviour *per se* in the everyday life of the people. Rather they should be viewed as a set of orderly, institutional arrangements at the societal level. Only at that level can one relate cultural behaviour to macroeconomic activities. Again, one is not anxious to find in these cultural areas the 'causes' of development: one can,

however, expect to look for the 'trigger' of development in East Asia (Hsiao, M.H.H., 1988, p. 20).

On the basis of available evidence, no causal connections between culture and economic growth in the region can be identified with any confidence. Wherever one looks in the West Pacific Rim countries, all the growth-promoting facets seem, in the final analysis, to be primarily structural rather than cultural:

> East Asian culture has remained largely constant over hundreds of years, yet the region's explosive economic growth is a matter of the past few decades. Culture provides a background of values reinforcing respect for authority, providing education, and rewarding diligence. But without the appropriate political structures, such a background will not produce results (Hofheinz and Calder, 1982, p. 250).

If we accept this conclusion, then it means that we can rule out the idea of a cultural model, or the search for particular combinations of cultural factors, characteristics and circumstances in order to explain differential levels of development. Without in any sense relegating culture to 'the dustbin of developmental economics' (Riedel, 1988, p. 26), the evidence suggests that we should reject any crude cultural determinism, just as we have already rejected both crude geographical and historical determinism. For some, this goes too far, for 'culture can either help or hinder development – it is up to the institutions, incentives and policies to determine which strands are allowed to dominate' (Gourevitch, 1989, p. 12). Yet such a moderate statement still seems to avoid outright rejection of cultural determinism. The experience of the West Pacific Rim countries suggests a firmer conclusion: that economic growth derives from an overpowering determination to achieve economic growth; that a conscious decision is made to pursue such growth; and that any cultural factors are simply used or rejected as seems appropriate.

4

The Chinese diaspora

Previous chapters have referred to the large numbers of Chinese to be found outside China in many parts of the West Pacific Rim. The point has already been made that they have generally achieved a much higher economic profile than their numbers would suggest. Furthermore, several very substantial studies have been made of the Chinese living outside China, especially in Southeast Asia (Coppel, 1983; Cushman and Wang Gungwu, 1988; Purcell, 1952, 1965; Suryadinata, 1989). This is hardly surprising. The Chinese exert a pervasive influence over so much of our area; indeed, one author argues that the main link between the various countries, cities and islands of the West Pacific Rim is the presence of the Overseas Chinese, 'the greatest diaspora in the world' (Winchester, 1991, p. 220).

While Chinese communities exist in Korea and Japan, their numbers and significance in those countries are not typical of the Chinese diaspora. Hong Kong and Taiwan are both peopled by what are strictly speaking Overseas Chinese, but in both cases Chinese make up almost the entire population and so are best considered separately. For the purpose of this chapter, therefore, the discussion will focus largely on the Chinese diaspora as it affects five countries: Malaysia, which includes the North Borneo territories of Sarawak and Sabah; Brunei; Singapore; Indonesia; and the Philippines.

Numbers of Overseas Chinese

As shown in Table 4.1, the numbers of Chinese and their proportions of the total populations in the countries of the West Pacific Rim today vary widely. It is difficult to be certain about these figures or to get hold of precise information about the role played by the Chinese in economic development. All one can say with some confidence is that the figures and information are probably more reliable for Singapore and Malaysia than for elsewhere. One difficulty is the definition of 'Chinese'. In the Philippines, for example, many Chinese, both ethnic Chinese (pure Chinese) and *mestizos* (a mixture of Chinese and some other race) have adopted Filipino names and Catholicism; their children are educated in Filipino schools; they speak one or more of the many Filipino languages, even at home; and they have taken on Filipino nationality. In most respects they are Filipino rather than Chinese. To a lesser extent the same problem arises in Malaysia, where what are known as the *Baba* Chinese in the southwest of the peninsula have adopted Malay dress, eat Malay-type food, and often speak Malay.

Table 4.1 Numbers of Chinese in Southeast Asia

	Ethnic Chinese	Total population	% Chinese
Indonesia	4,116,000	147,000,000	2.8
Philippines	699,000	46,000,000	1.5
Malaysia	4,882,300	12,736,637	30.9
Singapore	2,038,000	2,413,945	75.9
Brunei	85,000	300,000	28.0

Source: Various, but especially Rigg (1991) p. 110, and Suryadinata (1989) p. 1.

Another complicating factor is that the Chinese are not a clearly defined and homogeneous group. The Overseas Chinese come from many different parts of China, though mostly from southeastern China. They speak different dialects, most of which are mutually unintelligible, and they have different beliefs, values and predilections. The Hokkien, Cantonese, Hakka, Teochew, Hainanese and many other groups do not always operate together as one might expect; and they often specialise in their economic activities.

Nevertheless, the figures show that in Indonesia there are over 4 million Chinese, or about 3 per cent of the total population. In Malaysia there are rather more, almost 5 million Chinese, but this gives them a much higher percentage of 30 to 35 per cent of the total population. Brunei has a Chinese percentage of almost 30 per cent, and in the Philippines fewer than a million Chinese represent a percentage of the total population of about 1.5 per cent. Singapore has easily the largest percentage (76 per cent) of Chinese. In the five countries under discussion here there are altogether some 12 million Chinese, though the total number of Overseas Chinese in the West Pacific Rim, including Hong Kong, Taiwan and Vietnam, is about 40 million.

Origins of the diaspora

A great deal has been written on the origins of the spread of Chinese into Southeast Asia (Hall, 1985; Purcell, 1952, 1965). It is a long and fascinating story, for Chinese intellectual, cultural and commercial influence goes back at least to the third century AD, when official missions were despatched to report on countries bordering *Nanyang* (the South Seas), to be followed by Buddhist pilgrims and later, during the Sung period, by traders. After conquering China the Mongols also traded in the South Seas and with Arab traders in the area. This sequence of contacts 'explains why the attitudes and sympathies of these mercantile classes underlie all Overseas Chinese life today' (Winchester, 1991, p. 238).

Our interest here, however, is briefly with why and how the Chinese diaspora took place. All the evidence suggests that the primary motive was economic. Most of the migrants came from the heavily populated provinces of southeastern China – Guangdong, Fujian and Guangxi – where poverty and recurrent famines were common. There were clear economic forces tending to push the Chinese out of their homelands. But there was also a good deal of internal strife and physical insecurity, and it is a matter of interpretation whether the economic 'push' motive was as clear-cut as is commonly suggested. There were certainly very strong economic 'pull' factors encouraging the Chinese to look overseas for a better life.

The economic effects of the Chinese in Southeast Asia were felt well before the colonial period began: 'Chinese trade, exploitation of resources and introduction of technology made a major contribution to the region's economies long before the nineteenth century' (Dixon, 1991, p. 45). But it was during the European colonial period that the most marked and sustained movement of Chinese into Southeast Asia took place. At first the movement was largely voluntary, but as the need for labour grew, especially in the mines and plantations of the European colonies, the Chinese were recruited more formally and on an increasing scale. In pre-colonial Malaya there were already substantial numbers of Chinese engaged in a wide range of activities, including trade and some mining of tin. But when the British came and began to develop mining and rubber plantations, the Chinese were encouraged to arrive in much larger numbers, just as Indians (Tamils) were later encouraged to come in to work the rubber plantations. Numerically, the Chinese soon came to dominate the major urban centres. And in Malaya under the British, in the Philippines under the Spanish (and later under the Americans), and in Indonesia under the Dutch, the Chinese soon made themselves indispensable as traders, entrepreneurs, artisans and food producers.

As noted in Chapter two, some writers have suggested that the

Overseas Chinese communities were used by the colonial powers deliberately to suppress the indigenous populations and to link the populations and the resources of their countries into a growing and insatiable world capitalist system. The Chinese formed the middlemen – a link, or a channel of exploitation – between the indigenous populations and the colonial rulers. Whatever the truth of this matter, the Chinese were undoubtedly regarded as a real asset by the Europeans in their colonising process. The numbers and economic significance of the Chinese grew during the colonial period, restrictions on their immigration only being imposed with the advent of the Depression after 1929; in the Philippines, however, earlier and stricter controls were imposed by the Americans.

It is important to remember that the Chinese originated from a society that was in many ways much more sophisticated than they found in the countries in which they settled: they had a relatively developed technology and a long history of urbanisation, trade and administration. On the other hand, when the Europeans arrived they found that the indigenous population had little interest or experience in such matters as administration, supervisory work, or mining; in trading or trading services; or in supplying goods to more remote or newly developed areas. The Chinese seemed to be far more enterprising and hard-working and, being outsiders, they cared little for changing political frontiers or for traditional indigenous or religious differences. This, at least, is the conventional interpretation, reflecting the kind of attitude towards the indigenous peoples taken by Boeke (1966) and referred to in the previous chapter. It certainly appears that colonial administrators valued the Chinese in their territories. Fisher (1971), too, notes that the colonial powers thought the indigenous peoples in the region unsuited to modern economic activity and, moreover, that they were unwilling to give up their traditional ways of life in order to take part in the hard work demanded by Europeans. Furthermore, he points out that the colonialists were operating in areas with low population densities and that it was generally agreed that mine, plantation and other labour could only be found through a policy of encouraging immigrant labour.

However, the Chinese, despite their usefulness, were regarded with some ambivalence by the European colonial powers. This had for long been true in the Philippines under the Spanish, who became more and more concerned at the growing number of Chinese and their increasing economic strength; and this concern continued after the Americans took over the territory in 1895. Nevertheless the Chinese did help to promote commerce in the Philippines. By the end of Spanish rule there were over 100,000 Chinese residents. Growing world trade during the nineteenth century and the introduction of steamships meant wider opportunities for growing export crops, such as sugar, hemp, tobacco, coffee and copra; in all this development the Chinese played a key role.

In Malaysia, too, similarly ambivalent attitudes towards the Chinese were evident. The British initially encouraged Chinese immigration and this remained largely unrestricted until 1929. Many of the Chinese came to mine tin on the peninsula and gold in Sarawak. In 1920, 64 per cent of all tin mines in Malaya were owned by Chinese, though as the Europeans introduced their advanced technology the Chinese gradually lost their dominance. British attitudes towards the Chinese were in some ways quite openly discriminatory. Malays were given preference in such activities as rice farming; land was reserved for Malays; the Malays were given preference in the civil service; and when restrictions on the immigration of Chinese came into effect some voluntary repatriation also took place.

Economic significance

It is beyond question that the economic significance of the Overseas Chinese in the countries of the West Pacific Rim remains very considerable, and some would argue that the Overseas Chinese in Southeast Asia have played and continue to play a crucial role in providing the catalyst for economic growth in the region. As Redding (1990) points out, the 40 million or so Overseas Chinese in the area have between them a GNP two-thirds the size of the 1.1 billion Chinese in China. In peninsular Malaysia the Chinese own and control nearly 40 per cent of the corporate sector and their contribution to economic development has been critical. In Sarawak, too, where most of the non-indigenous people are Chinese, they occupy the more densely populated coastal and lower valley strips in the relatively well-developed west. The Chinese, too, are the main urban people in Sarawak, as they are throughout most of Malaysia: 60 per cent of the population of Kuching, the capital of Sarawak, is Chinese. Throughout Malaysia the Chinese dominate in business and it is generally true that the whole internal trade of the country passes at some stage or other through the hands of Chinese merchants or middlemen.

The same is largely true of Indonesia, where Chinese dominate in most commercial sectors. Chinese operate especially at the higher levels of trade, capital-intensive and high-technology trade being mostly in Chinese hands. Property markets are also dominated by the Chinese; Chinese private firms are growing rapidly; and Chinese companies dominate in shipping. In the Philippines, Chinese own some 40 per cent of the total assets of private domestic banks.

Clearly the Chinese are an important element in the economic life of these Southeast Asian countries. Explanations of their success are many and varied: 'the reasons used to explain the success of the Chinese in the region include their high motivation (which is linked to their status as migrants), their Confucian work ethic and business acumen, and the role of Chinese business networks' (Rigg, 1991, p. 111). They also include the suggestion that the Chinese happened to be in the right place at the right time; and that they had the skills, the temperament, drive, experience and urbanised culture that coincided with the needs of the European colonialists. Moreover, as noted in the previous chapter, the Chinese are said to be hard-working, frugal, patient and willing to make less profit per unit sale. Their work ethic is certainly very strong and sometimes frighteningly uncompromising. The stigma attached to play and relaxation is also very apparent, especially among the older Chinese. It is also pointed out that the Chinese have a propensity to invest accumulated capital in a trade in which the entrepreneur has acquired certain basic skills through previous apprenticeship or in-service training as an employee.

Explanations of why Chinese family businesses in Hong Kong, Taiwan and Southeast Asia have been so successful are also usually largely cultural. Redding, for instance, argues that it is culture – the economic roles and effects of family structure, filial piety and work ethic – that keeps them small, compared with other capitalist economies; he also puts some weight on their faith in prophecy (Redding, 1990).

However, two other explanations have received rather less attention in the literature. The first is the Chinese attitude to kinship. This is important for employment, in establishing trade and in its subsequent expansion; and it has been the most important cause of dialect trade specialisation and dominance. The second is that the family organisation from which most businesses are formed and expand is based on personal credit standing, relationships and effort.

In their attitude to kinship, there are important social structures, other than the family, that play a key role in Chinese business organisation. Traditionally, the various dialect groups were structured along the lines of *bang* ('a grouping' or 'a gang'). *Bang* is based primarily on dialect grouping, perhaps best illustrated in Singapore (Cheng, 1985). The Singapore Chinese Chamber of Commerce and Industry (SCCCI), established in 1906, was in fact based on *bang*. Although common economic interests had brought the leaders of different *bang* together, the lack of a common tongue – the various dialects being to a large extent mutually unintelligible – and inter-*bang* rivalry tended to keep them apart. In order to prevent the proposed association from being dominated by a particular *bang*, a *bang*-structured SCCCI was created.

The functions of *bang* were social, political and economic. The SCCCI acted as a forum for communication between *bang*, and this was very necessary, for they often acted independently to compete with and exclude each other, especially as far as economic activities were concerned; and each *bang* tended to specialise in and effectively monopolise particular areas of trade. The members of the specialised trade associations in each *bang* were encouraged to take over any of its members' firms that were in danger of being wound up. One consequence of this dialect trade specialisation was therefore that the particular dialect became the language of that trade. Dialect patronage and trade associations are mutually influencing and reinforcing: they form a barrier by excluding members of other dialect groups from gaining entry into that trade or at least from being effective in it.

However, as Cheng (1985) points out, *bang* associations are declining in size and importance as development proceeds, more especially in a society where multicultural policies are being pursued so energetically, and where recent educational, economic and socio-cultural changes make the *raison d'être* of *bang* less easy to defend.

As for personal credit standing, or *xinyong*, there can be no business on credit without this feature. Business transactions over the phone involving deals worth millions of dollars are said to be common. It is a verbal agreement, without any need for legal contract; but, once made, it must be honoured. News of reneging on a deal spreads quickly and can lead to severe sanctions being imposed against a trader. Arbitration on bad debts may occur through the trade associations and *bang*. Good credit standing extends into good personal relations and what is known as *ganqing* (affection) and this affects good personal contacts in business.

A further characteristic of the Chinese which affects profoundly the operation of Chinese companies throughout the countries of the West Pacific Rim is *guanxi*. This refers to connections between members of the same family, village, clan, or simply with ancestors from the same province. *Guanxi* networks are global, as are the business companies, and they provide the essential links by which Chinese entrepreneurs can operate efficiently over huge distances.

The origins of all these characteristics are impossible to determine precisely. The Chinese historian Ho Ping-ti, a highly respected Chinese scholar, has argued that three factors contributed to the formation of what he calls strong locality-consciousness and thus to dialect trade specialisation and *bang*. First, Confucian codes of ethics and laws; second, administrative regulations concerning the appointment of government officials on the basis of geographical origin; and third, the system of Imperial Civil Service Examinations. The early associations (*Lui Kuan*) that appeared in Beijing were formed by officials stationed in Beijing and were based on the principle of common geographical origin. Later, these associations developed to accommodate traders and candidates for the Imperial Civil Service Examination. Ho argues that the establishment of similar associations outside by the Overseas Chinese in

Southeast Asia represents the transplanting of this Chinese tradition overseas (Ho, 1966).

However, locality-consciousness is not confined to the Chinese. It is found among all immigrants in all parts of the world, and it is inevitable that, if dialects are mutually unintelligible and ways of behaviour and even local gods are different, these groups will band together. Confucianism and administrative systems seem unlikely to have had much effect; locality-consciousness seems more likely to have arisen quite simply out of a natural tendency towards localism.

Whatever the mechanisms used by the Chinese, whatever their origins, and whatever the reasons for their adeptness at trade, these aptitudes of the Chinese were understandably encouraged and fostered by the Europeans throughout Southeast Asia. It was perhaps inevitable that the Chinese, being outsiders, would find themselves in the position of middlemen and would feel that they had to excel at whatever they did. The Chinese were also in most cases not allowed by the colonial governments to participate in other activities, such as politics or in the professions.

That the Chinese have made an important, even critical, contribution to economic growth in Singapore, Malaysia, Brunei, Indonesia and the Philippines is undeniable. But at the same time they have had to operate within the societies and political systems of the host countries, and it is arguable whether they have been entirely successful in these fields. One way of discussing this matter is to look briefly at the reactions to the Overseas Chinese of, first, the indigenous people in the host countries and, secondly, the Chinese in mainland China.

Reactions to the Overseas Chinese

The reactions of the indigenous peoples to the presence of Overseas Chinese in their midst is a particularly sensitive issue because it carries with it a number of racial overtones. Nevertheless it is certainly true that the common view among the local people of Malaysia, Indonesia and the Philippines is that the Chinese are exclusive, cohesive, not prepared to integrate; that they have developed and still maintain a stranglehold on economic life; and that their cohesion, economic strength and strong sense of cultural identity make them sympathetic to the interests of other Chinese, from wherever they come, rather than to the interests of the host country.

In the Philippines, for instance, it is widely believed that the Chinese form a separate cohesive block; that there is ethnic and linguistic homogeneity among the Chinese; that they possess strong leadership; and that mixed marriages are rare. The schools the Chinese send their children to are Chinese schools which inculcate Chinese religions and political attitudes inimical to the interests of the host nation. The Chinese also have their own newspapers and other printed material which 'spread sedition' and are used to retain and develop economic control. Such views are reinforced by the existence of Chinatowns – a ubiquitous feature in Southeast Asia – in even the smaller urban centres in the Philippines.

Yet there is probably little truth in most of these perceptions. As already indicated, the Chinese are by no means a homogeneous group: they comprise many different groups distinguished by dialect, religion, business and clan orientation; and of course the family is in any case the overriding unit. Some writers argue that in fact there is no communal cohesiveness, let alone a common political ideology, a common religion or any strong, unified leadership. This is indeed hardly surprising, considering that the Chinese are composed of many different groups from different parts of China. After all, China itself is a vast and highly fragmented country where transport has always been and remains difficult; a country in which it is often hard to believe that there really is an outside world; a country in which different people live according to different rules and seemingly in different time-scales. The Chinese on the mainland speak what are in effect different languages over quite small distances. Thus Shanghai has its own language, as does Suzhou and Nanjing, all within a few hundred kilometres of each other. There is therefore a great deal of parochialism and a sense of being isolated from the outside world. China is a country where one feels that everything is moving slowly: a claustrophobic, strangely unreal and anachronistic country.

Even where most of the Chinese seem to originate from the same dialect group, as in the Philippines, where most of the Chinese are Hokkien, there are many subdivisions: different clan associations with separate community halls, customs and ideologies; associations based on the provinces, districts and prefectures from which the Hokkien originate; small district and village associations; and even tiny associations based on family groups. In the Philippines, even Chinese schools reflect differing clan associations; they cannot be said to cultivate a common sense of belonging to some monolithic Chinese culture.

This fragmented nature of the Chinese is also exhibited in the apparently cohesive Chinatowns, where different groups often concentrate for business and residence in a particular area. Certainly in the Philippines the Chinatowns are rarely exclusively Chinese, containing as they do an increasing number of Filipino-run businesses. Moreover, the composition of the Chinatowns is continually changing. Once they have the means, the Chinese prefer to live among the indigenous

Filipinos, and in Manila there is a clear tendency for the Chinese to move out of Chinatown to the surrounding suburbs.

The notions that mixed marriages are rare, that Chinese newspapers and schools are run for seditious purposes and for binding the Chinese tightly together so as to exclude the indigenous population, all these are again based on emotion rather than fact. As noted earlier, in the Philippines the population of *mestizos* – Chinese/Malay/ Filipino and/or European – is large, though there are no reliable published figures. Chinese newspapers are used to make known reports carried in the Philippine media – international, cultural and business news – to *all* Chinese, the Chinese script being the means by which Chinese from different dialect groups communicate with one another.

Nevertheless, there is no denying that there is a good deal of anti-Chinese feeling in the host countries. Economic nationalism and political discrimination are widespread and, as we shall note in a later chapter, perhaps the most difficult case in the region is Malaysia, where the proportion of Chinese is approaching a third of the total population. Countless reasons can be advanced to explain this antagonism. But there is little doubt that, underlying it all, is the relative economic success of the Chinese.

Indonesia provides a good case in point. In the late 1950s the economic nationalists, those who objected to the economic role of the Chinese, attempted to undermine Chinese economic power. The Chinese were forbidden to engage in retail trade in rural areas, and various other measures were taken to prevent the rise of Chinese businesses. But this soon caused economic problems for the Sukarno regime itself. According to one writer, 'long years of neglect and mismanagement, aggravated by the persistent refusal of the Indonesians to make anything like full use of the administrative and commercial talents of the Eurasian and Chinese communities, had brought the economy to the point of collapse by 1965' (Fisher, 1971, p. 303).

The Suharto regime which followed the downfall of Sukarno introduced a more liberal economic policy and massive foreign investment was welcomed. Some writers maintain that the Chinese are still excluded from several key sectors of the economy, such as oil and minerals and some of the export trade where large state companies dominate. However, restrictions on the Chinese have been gradually released and large and very powerful *Zhugong* have developed. This term describes a business concern where a Chinese businessman collaborates with a member of the Indonesian power élite; it is an informal arrangement and carries with it overtones of what Western observers might interpret as corruption. The Chinese *Zhugong* provides the skills and sometimes the capital in running the business, while the Indonesian gives protection and facilities. Some authors suggest that the government has deliberately encouraged the growth of *Zhugong* and the general involvement of the Chinese in business. It is a useful arrangement for the power élite; it is good for the Chinese; it is good for the country's economic growth; it encourages the growth of a strong indigenous middle class; and it gives the government more control over the politically weak and vulnerable Chinese. There have been problems, of course. Tensions have arisen between the ethnic and Indonesian Chinese and the Indonesian businessmen – the *Pribmi* – often supported by Islamic groups. Nevertheless the Chinese *Zhugong* acquired considerable importance and many have developed into large multinational corporations.

Any examination of anti-Chinese sentiments in the host countries must acknowledge that factional divisions of all kinds are endemic in the region. Such factions include other ethnic and religious groups and their many subdivisions. They include also those groups often given the easy and, because it has so many different meanings, the rather abstract and meaningless term 'élite'. The divisions between these élite groups may follow ethnic or racial lines, but there are many subdivisions and non-ethnic divisions within the Chinese, other immigrant and the indigenous populations.

A particularly good example is the Philippines where the numbers of Chinese are small, but where factionalism is almost a way of life (Mackie, 1988, p. 325). There are perhaps ninety to one hundred or more different indigenous languages and dialects, many of them mutually unintelligible (Alip, 1974; Blaker, 1970), and political groupings often draw upon these linguistic–cultural divisions. For instance, Ferdinand Marcos drew upon the Ilocano, Imelda Marcos drew her support from the Waray or Leytonias; and Corazón Aquino drew her support from the Kapampangan of Pampangga province. These divisions are far more sophisticated and subtle than is sometimes suggested. These political, economic, linguistic and cultural groups draw upon the extended family, which is often remarkably extensive, and upon a complex and shifting network of personal contacts: friends, people who owe favours, and people who are owed favours as well. All these elements combine with individual idiosyncracies, ideologies, political beliefs and economic circumstances both to divide and to unite.

Turning now to the reaction of mainland China, the attitude of the Beijing government towards the Overseas Chinese in some Southeast Asian states has occasionally given cause for legitimate concern. But most often, perhaps, anxiety over the links between the Overseas Chinese and mainland China has simply provided a convenient political tool for the governments of Southeast Asian states.

The Overseas Chinese – the *Huaqiao* – have been important to the People's Republic of China, both politically and economically. But mainland China's attitude towards the Overseas Chinese has been ambivalent, for trusting the Overseas Chinese does not come easily to Beijing. After all, many of them left China to get away from Communism and nearly all are regarded as thoroughly committed capitalists. But Beijing's concern over the condition of the Overseas Chinese is to a large extent a propaganda exercise, promoting a non-existent uniform Chinese culture.

In Indonesia, as we have already seen, the Indonesian government in 1958 and 1960 prohibited its Chinese population from engaging in rural retail trade and even from living anywhere in West Java. Beijing intervened in this decision, maintaining that the Sukarno government of Indonesia had violated the Dual Nationality Agreement. In China a campaign was launched against Sukarno, and Beijing even sent ships to expatriate the Overseas Chinese. However, when military aid to Indonesia from the Soviet Union increased, China discontinued the dispute. Even when many Chinese were killed in Indonesia during the 1964 anti-Chinese riots, Beijing blamed abstract 'reactionary elements' rather than Sukarno. Later, the 1965 failed coup in Indonesia brought into power a more anti-Beijing government. An anti-ethnic Chinese campaign was launched in Indonesia, and again mainland China sent ships to expatriate the Overseas Chinese. Yet when in the 1970s the restoration of relations with the Indonesian government became possible, Beijing turned a blind eye to sporadic anti-Chinese riots; by then it was in both governments' interests to re-establish good relations. From the point of view of mainland China, the Overseas Chinese are now largely a lost cause. Some of them may have been Communist Party members, and a small percentage were no doubt engaged in subversive activities. But they were never seriously regarded as providing an entrée into the internal affairs of another country. In the late 1970s Deng Xiaoping made it clear that China was prepared to recognise reality: many Overseas Chinese do not want to be referred to as Overseas Chinese or indeed to be connected with mainland China in any way.

Conclusions

That divisions exist and tensions arise between groups or factions, and that these may either hinder or promote economic development, is clear. But this should not obscure the fact that the Overseas Chinese continue to play a critical role in the economic development of those countries in which they now live. As entrepreneurs, in commerce and in business, the Overseas Chinese continue to act as the catalyst of economic growth. They have been described as the 'omnipresent, inescapable part of the warp and woof of every country between Vietnam and India, Luzon and Timor' (Winchester, 1991, p. 237).

5

Exchange and domestic trade

The previous chapter referred to the important entrepreneurial skills of the Overseas Chinese in the West Pacific Rim and to the growth of commerce on which so much of the economic prosperity of the region has been and must continue to be based. Attempts in the present chapter to examine these matters in rather more detail, however, come up against the remarkable lack of relevant literature. While there is plenty of material on foreign or international trade (to be discussed in the next chapter) there is very little on domestic trade or commercial development. To some extent this reflects the fact that useful material on domestic trade and commerce almost always demands the collection of field data over quite small and not necessarily representative areas. Most countries have very few or no published works on which one can draw.

This chapter focuses on the case of China, on which a certain amount of important work on domestic trade and commercial development has been published, particularly by Skinner (1964–65; 1985) and Solinger (1984). The main reason for choosing to focus on China, however, is that it provides a very clear illustration of what happens if trade is seriously interfered with. The discussion examines particularly the changes that have been taking place in China since the Reforms of 1978, changes which have contributed significantly to the substantial economic progress of the country since that year. It is true that the degree of state control in domestic trade and commercial development is, admittedly, still greater than in probably any other country. But present changes in the balance between private and state domestic trade and commercial development do at least facilitate an assessment of the relative contributions of each to the process of economic growth. More importantly, they highlight the fact that, if the logic of the reforms in China 'is to be pursued consistently, and if reform policies are to achieve their desired effects across the board, comprehensive changes in commodity circulation and exchange are necessary' (White, 1988a, p.186).

Difficulties in analysing the development of private as distinct from state-controlled trade in China since 1978 arise not only from the inevitable dearth of reliable statistics and other information. There is also, in the Chinese-language literature, a reluctance to engage in an open debate on the development of private trade broad enough to include a thorough and balanced consideration of its economic and political significance. This reluctance reflects to some extent the long-standing ideological objections to private trade in China; and this denial or marginalisation of private trade's economic importance can be construed as an attempt in certain quarters within China's administration to conceal its potential value and to re-emphasise the continuing relevance and efficacy of public ownership and the state system of procurement and distribution. Thus the potential contribution of private trade to the economy is underplayed and its negative effects exaggerated. At the same time, the administration and academics in China are only too well aware that the development of private trade in the country has become, especially in the last few years, a politically sensitive issue. Yet without a dynamic and increasingly sophisticated system of domestic trade the recent economic improvements will inevitably be constrained.

Markets and trade networks in China

This is as true of market-place trade as it is of the broader field of domestic trade in China. The literature on the origins, growth and characteristics of local markets in China is dominated by the classic work of Skinner (1964–65; 1985), who shows how critical these phenomena are for any understanding of domestic trade in China. In his 1985 reassessment of market-place trade Skinner emphasised that in the pre-1978 period markets were never really free markets, and that their very existence was to some extent an affront to the government's economic philosophy:

> Rural markets and peasant marketing did not fare well during the Maoist era . . . Maoist radicals . . . can be fairly characterised as having an anti-market mentality. While this set of attitudes derives in part from Marxism, it is also rooted in the ideological preconceptions of late-imperial Confucian bureaucrats. (1985, p. 393)

He goes on to state that opposition to the market stemmed from the fact that the market mechanism disrupts communities and diminishes collectivity; markets mean haggling and introduce competition; the market reduces motivation down to the family level and flouts self-sufficiency;

and the market-place attracts 'ne'r do wells and breeds "shysters" and bullies' (Skinner, 1985, p. 393). In other words, service to the community is ill-served by market exchange. As a result, the Maoists, especially after 1953, implemented a programme that diverted rural trade to supply and marketing co-operatives and state trading companies, reduced the number of markets, and sharply curtailed marketing activity.

Since 1978, however, the encouragement of economic autonomy for persons and households throughout rural China has created an institutional framework and incentive structure that favour entrepreneurial initiative and improved productivity, especially in agriculture. The revival of rural marketing is a critical element in these reforms. It was agreed that rural markets should not be considered capitalistic. Yet while rural markets were thereby legitimised, the government retained virtually all the trade controls of the earlier phases, at least until free markets were allowed to take place within municipal boundaries. Subsequently, further relaxations occurred and by 1984 rural markets were less restricted than at any time since 1953. As a result the number of rural markets increased dramatically – by over 30 per cent in the five years from 1978 to 1983. Furthermore, there was an increase in the frequency of periodicity and an increase in the amount of trade turnover in the markets (Skinner, 1985, pp 407–8).

An important concomitant of this increase was the removal of restrictions on long-distance trade after 1979. This, according to Skinner, has led to the re-emergence of a marketing hierarchy.

> Current policies aim to rationalise economic networks so that commodities may be transferred from production sites to the point of consumption 'along the shortest route with the fewest links and at the lowest cost', and to 'integrate the economic activities in large, medium-sized and small cities of various types with the economic activities in rural townships'. (1985, p. 410).

While periodic marketing is likely to decline in China, this is to be expected and indeed welcomed in that, as research elsewhere has shown, an increase in periodicity will lead ineluctably to daily marketing and eventually to shops and to shopping and commercial centres. Certainly in China the number of 'free' markets is expanding, though a growing number of them are no longer either traditional or periodic. They have permanent buildings and facilities and are open daily. Skinner notes that 'the dramatic revival of marketing is integral to the larger liberating reform of the rural economy – both responding to and stimulating in turn the increases in specialization and productivity' (Skinner, 1985, p. 412).

The growth of private trade in China

In the attempt to create a market economy, albeit a 'socialist market economy', in China, the further development of private trade is an essential prerequisite. Since the Reforms of 1978 the growth in the number of private trade organisations and in retail sales by private traders in China has been dramatic. The number of private trade organisations rose from about 100,000 in 1978 to over fourteen million by the late 1980s; and over the same period the private traders' share of total retail sales increased from under 0.5 per cent to about 25 per cent. Private traders are now regarded by the Chinese state administration as essential, though still largely supplementary to the state and collective systems of ownership and, in particular, to those state-run and collective-run units which comprise the state system of procurement and distribution.

The need for a thriving system of private trade was recognised in 1978, by which time the state system had become cloistered and rigid. The logistics of arranging and operating the existing vast network of state procurement and distribution required fragmented administrative responsibilities (and therefore fragmented economic activities) held together and coordinated by vertically structured administrative command routes. Self-sufficiency and the creation of defensive administrative units were natural accompaniments, and both phenomena were enhanced by the need to reduce the physical and fiscal burdens on the state system of procurement and distribution. These administrative units and their affiliated economic enterprises, vertically integrated through administrative command routes, provided in effect no mechanism for the creation of wealth; under this system there could be no interaction or exchange between independent parties.

It was recognised that the creation of truly independent economic organisations would give rise to an increasingly atomistic economy in which organisations and individuals would have to depend largely upon the exchange *for profit* of goods, skills and information. In short, the administrative fabric, and in particular, the state system of procurement and distribution which had previously bound economic organisations and individuals together, would be replaced by *trade*, i.e. by exchange for profit.

There were difficulties, however. Such a course of action was logically incompatible with the maintenance and predominance of the state and collective operations. The state has tried to imitate an economy dominated by trade relations, while at the same time refusing to allow the degree of economic freedom which lies at the heart of any successful market economy. Attempts have been made to devolve power to economic organisations: there has been an attempt to draw a clear distinction between profits and taxes; some industrial enterprises have been allowed to develop their own wholesale and retail outlets; horizontal linkages among many economic organisations have been established, irrespective of administrative divisions (Hodder, 1990); and more flexible forms of procurement and distribution have been introduced. To some extent state-run and collective-run units have participated in trade as they become entwined with the operations of private traders. But although the development of private trade has in this sense been permitted, the state has required private trade to remain supplemental to the production, procurement and distribution of goods and materials by the state's administrative organs and their affiliated economic units.

Evidently the state's attitude towards private trade has been and remains somewhat ambivalent. Private trade is a cardinal element of the Reforms; and yet it must remain subordinate to the state system of procurement and distribution. This ambivalence betrays a long-standing conflict between state bureaucracy and merchants. At the heart of this conflict lies a tension between the two forms of exchange – private trade, and the state system of procurement and distribution – and their associated political, strategic and economic imperatives. Moreover, in 1989 – possibly as a reaction to the feelings which led up to and followed the events in Tiananmen Square in Beijing in June – the official attitude towards private trade hardened. Despite its relatively weak position within the economy as a whole, private trade was now perceived to be capable of eroding the fundamental elements of state control. Private trade directly challenged and actively weakened the state-run and collective-run enterprises.

In addition, the initially rapid and successful development of private trade after 1978 added a disturbing psychological dimension to the gradual shift away from the emphasis previously given to constant increases in production and productivity. Some authorities took the view that difficulties with the provision of goods and materials were essentially problems of supply, and that increases in production and productivity were a natural spin-off from improvements in the circulation of goods in demand. Far more important, however, was the emergence of the philosophy that wealth creation is rooted in exchange for profit between economic organisations. Private trade and the wealth it generated was a sudden corporate expression of an unvoiced and intuitive understanding of the importance of trade and the benefits it could bring to every individual. By implication, private trade questioned the need for the individual to subordinate his interests to the collective efforts of state and society in order to increase output; and, by definition, private trade demanded the separation of economic units from state administrative organs.

Private trade developed independently of the state, but it quickly became entwined with state-run and collective-run economic units and began to pull these units away from the administration. This insidious process was made easier by the emphasis which the state itself had given to commercial exchange and to the loosening of administrative control, as part of the attempt to imitate a system of private ownership and to engineer a commercially integrated economy. The devolution of greater power to economic organisations, coupled with the rapid growth of private trade, came to be seen in some quarters as tacit encouragement of *de facto* privatisation, and was said to be bringing about the gradual disintegration of the state and collective system of ownership. In particular, the contracting and subcontracting of many of those units comprising the system of procurement and distribution and its infiltration by private trade had led to parasitism and inertia, and had so weakened the fabric of that system that it had begun to crumble.

If left unchecked, so it was argued in the late 1980s, the logical conclusion of the development of private trade would be the fragmentation and collapse of the socialist organisation of the state. Economic units would become independent. Indeed, there were those who argued quite openly that the gradual disassociation of economic organisations from state administration would have to be followed by privatisation. By extension, independent economic organisations would bring about the evolution of a highly atomistic and therefore, by necessity, an interdependent society.

There were already worrying signs of this. Narcissism and the importance which many, especially the young, attached to 'self' was regarded as an indication of the formation of trade relations between individuals. These relations, selfishness and individualism, were perceived to be intimately linked with one another, and to have invaded the society and polity of China. Trade was a breeding ground for vice and political corruption; it attracted the dregs of society; and it provided the means and reinforced the values that could lead to organised dissent. The tensions generated had already exploded into assaults on tax offices and on tax officials; and counter-revolutionary and illegal organisations were infused with the values of individualism and 'self' and were therefore linked to the activities of traders. Indeed, there were rumours that private traders had helped to finance the pro-democracy movement which led to the Tiananmen Square massacre of 1989.

How far Tiananmen Square was the major cause of policy changes towards private trade in 1989 is a matter of opinion, though certainly outside China a simple causal connection is usually drawn. Perhaps, however, Tiananmen Square was only the specific excuse for such policy changes; perhaps the real causes were long-standing, deeply embedded within China's economy, society and polity. Nevertheless, the result was that the permanence and inviolability of the state and collective system of ownership were reasserted; individualism was denigrated and the place of the family within a much wider collective society was praised; and envy over the growing differences in income between many traders and the rest of the population was used by the state to bolster its own moral position. A life which made no contribution to society was to be regarded as a life without meaning. The aims of the individual were set out clearly: to contribute towards peace, stability, unity, modernisation and reform. More directly, private trade was to be restricted to a limited supplementary role; its share of retail sales, its influence on wholesaling, its profits and the type of activities in which traders could participate were all to be strictly controlled.

The importance and necessity of strengthening the state system of procurement and distribution was now given renewed emphasis. It was argued that the state system of procurement and distribution must remain the prime channel for the flow of goods and materials. After all, by monopolising the procurement, movement and distribution of goods and materials, the system had proved that it can guarantee civil and strategic (military and non-military) supplies; it allows tight control to be exercised over prices and the money supply; and it eliminates usury and the disruption of the domestic market by private traders. The state system thus acts as a crucial stabilising and unifying force, as it did during the disturbances of 1989 and at other times of political upheaval over the last forty years. The state system's ability to procure, move, store and distribute goods and materials, especially those categories which immediately and directly affect the general populace – agricultural and sideline goods and industrial consumer goods – has to be made secure in order to safeguard social and political stability. It is therefore of critical importance that: the state organs of procurement and distribution should be given direct and priority access to listed goods and materials; the system should be provided with loans at preferential rates; it should be managed more efficiently; administrative barriers should be removed; and private traders should be prevented from competing effectively with the state system of procurement and distribution.

Summarising the growth of private trade since 1978, it is clear that the authorities in China began attempting to modernise their economy through commercial integration. In part this attempt was a response to the rigid administrative structure that had evolved by the late 1970s. This structure had controlled and restricted economic affairs, had enhanced the tendency towards self-sufficiency, and had resulted in near stagnation in China's economy. The formation of a more successful commercially integrated economy clearly required that the administrative fabric which had bound economic organisations and individuals together should be replaced by exchange for profit between independent economic organisations and individuals. However, overriding politico-strategic considerations made it imperative that the basic administrative structure should remain intact. The authorities could only imitate an economy dominated by trade: vertical command routes and the state system of procurement and distribution were weakened; greater power was devolved to economic units; and the formation of horizontal linkages was encouraged.

The result was an intensification of competition for goods, materials, funds and markets; a strengthening of localism; and the duplication of economic activities. Moreover private trade, which had been permitted to develop only as a supplement to, and not instead of, state and collective systems of ownership, developed surprisingly fast. Private trade networks arose spontaneously and began to compete with and to infiltrate the state-owned and collective-owned sectors of the economy, including in particular the state system of procurement and distribution; and private trade began to invade – some argued to corrupt – the social fabric of China. It became clear that private trade was a powerful integrative force capable of eroding those very elements that enable the state to exercise control over the economy, society and polity of China.

By the late 1980s the authorities in China were thus faced with a dilemma. Either they had to accept and accommodate all that the development of private trade would mean for the institutions and power of the state, or they could subordinate economic matters to their concern for the politico-strategic implications of private trade, in which case they would have little choice but to restrict or suppress private trade altogether.

There are as yet no firm indications about how China will attempt to resolve this dilemma. It is at least possible that economic imperatives will dictate that private trade must be allowed to continue and to develop further, in which case private trade networks will continue to multiply rapidly, spontaneously and uncontrollably over the whole country, competing with and invading existing state systems of exchange. Or it may be that the state will attempt at least to confine the 'excesses' of private trade and its associated long-distance trading networks to those coastal parts of China where economic liberalisation has already resulted in dramatic improvements in economic prosperity. Yet while the government pays lip-service to the need for intra-China trade, including trade between the coast and the interior, this is unlikely to succeed without the necessary private trade networks.

The case of exchange and domestic trade in China is interesting for many reasons, and not only because it helps to demonstrate why the government there is ambivalent about allowing private exchange, markets and domestic trade to flourish as a necessary accompaniment to economic growth. It is also significant that whereas the Chinese are only now being allowed, however tentatively and reluctantly, to exercise their entrepreneurial and trading skills in China, the Overseas Chinese have long played a critical entrepreneurial and trading role in the domestic trade of most countries in Southeast Asia. As indicated briefly in the previous chapter, it is impossible to visit any country in the Southeast Asian sector of the West Pacific Rim without being made immediately aware of how crucially the domestic economies depend on the Chinese as traders, middlemen and businessmen. This is true even in those countries where the percentage of Chinese in the total population is small, as in Indonesia and the Philippines. It seems certain that without the 'age-old Chinese entrepreneurial instincts' described by Solinger (1984, p. 303), none of the countries in Southeast Asia would have been able to achieve their present levels of prosperity.

6

International trade and investment

It is now almost a truism that the major reason for the economic prosperity of the West Pacific Rim has been the region's involvement in international trade: 'Trade, trade and more trade was what propelled the Pacific Rim states out of agrarian destitution or post-World War II destruction and decline into world economic prominence' (Aikman, 1986, p. 10). As noted on a number of occasions in these pages, East Asia is generally believed to be the new emerging centre of gravity of the world economy. The area includes the world's most dynamic trading country and the four most successful Newly Industrialised Countries of Korea, Taiwan, Hong Kong and Singapore. During the late 1970s and early 1980s the region's countries increased their share of European and North American industrial markets, albeit from a low base and from a level which was small in absolute terms. Much of this trade was dominated by Japan, joined later by Hong Kong, Korea, Taiwan and Singapore, then by several of the ASEAN countries – notably Malaysia, Indonesia, and the Philippines – and finally, most recently, by China.

This achievement needs to be set against the conventional wisdom about the trading problems of developing countries, that their economic development has been restricted by existing patterns of foreign trade. Trade between Third World countries generally is said to be very limited: the range of their materials available for export is generally so small and so similar that they have very little to offer each other. International trade has also been controlled and governed by the metropolitan, former colonial, powers and has been directed largely at those crops and raw materials needed by the manufacturing nations of those powers and by the urban industrialised world. At independence, developing countries were also left with a legacy of external trading organisations which restricted rapid economic advance. In particular, the association between colonialism and export economies was emphasised by the concentration of the large-scale export trade in the hands of a few large, mostly European, firms and by multinational corporations.

A further problem faced by developing countries at independence, so the argument goes, was that their exports to the industrialised world were largely raw materials – in which they have a comparative advantage – while their imports were manufactured goods from the developed industrial countries. Primary, mainly agricultural products and some minerals, dominated exports, and finished consumer goods accounted for most imports. Furthermore, in many developing countries exports comprised only a very few items. Such a pattern of exports has repercussions upon the production side of the economy, on labour market policy, on soil fertility and conservation, on crop pests and diseases and on the improvement of dietary standards.

But the main problem usually highlighted in this kind of analysis is the economic instability caused by overdependence on one or two raw material exports. The purchasing power of such exports may fluctuate quite dramatically, especially where the manufacture of synthetics or substitutes is common and where demand patterns are continually changing. Then, too, many other technological developments are likely to work against the developing countries simply because they reduce the possibility of advance towards industrialisation by transforming exhaustible material assets into valuable man-made wealth; the whole tendency of technological advance is to make material resources more homogeneous, both in quality and in their distribution, and to reduce the actual or potential value of high quality natural resources that were once essential for industry. Again, increased agricultural productivity in the developed industrial countries has usually been accompanied by measures to protect those producers from adverse price effects arising from competition from low-income developing countries. Similarly, the protection of other domestic producers, such as oil and mineral producers, denies low-cost producers in the developing countries ready access to existing industrial markets. But whatever the causes of these fluctuations, they affect export earnings, balance of payments and customs dues. The problem, then, is that low-income countries today are typically faced with fluctuating and often declining world markets for their traditional raw materials. The pattern of world trade and the terms of trade appear to favour the industrial countries at the expense of low-income countries.

While it is true that several of the countries in the West Pacific Rim have in the past had some experience of many of these problems, this conventional interpretation no longer seems to have much validity. Indeed, as far as the countries of the West Pacific Rim are concerned, much of the general literature on developing countries seems increasingly to be out of date. Why this should be so is an interesting question. For at the end of the World War II the countries of the West Pacific Rim were all either in rural poverty or their economies had been devastated.

If, as is argued in this chapter, it was above all the growth of trade that subsequently produced such rapid economic progress, then how was this growth of trade achieved?

The most powerful stimulus and catalyst came from the United States. Determined to protect its own strategic interests against the eastward spread of communism and any resurgence of imperialism, the United States set about building up Japan's economy. Through massive aid and by opening up its own markets to Japanese industries, the United States very quickly set Japan upon a path of rapid economic advance. The prosperity built on trade after the 1950s in Japan and then in the four NICs depended very much on the openness of markets in the United States. In addition, as we have seen, the United States set up a triangular trading network between the United States, Japan and Southeast Asia, thereby diffusing the benefits of economic growth throughout the West Pacific Rim countries and providing an impressive example of how to achieve economic growth through international trade. It is true that Japan and the other territories in the region were, to differing extents, energetic and committed in their responses; that the indigenous entrepreneurial skills, notably of the Chinese spread throughout the region, were important; and that the insular and peninsular countries of the West Pacific Rim possess crucial geographical advantages for trade. But it was initially the self-interested actions and support of the United States that made it all possible.

Growth of trade

Whatever the reasons, the growth of international trade in the countries of the West Pacific Rim has been remarkable. From 1982 to 1988 the growth of export volume from East Asia (even excluding Japan) was just over 12 per cent per annum, a rate almost double that of South Asia, three times that of the Middle East, North Africa and Latin America, and about six times higher than in sub-Saharan Africa. At the same time the growth of real GNP in East Asia and Southeast Asia (again excluding Japan) was double that of South Asia, four times that of the Middle East, North Africa, Latin America and the Caribbean, and some eight times that of sub-Saharan Africa. Table 6.1 shows the growth of merchandise trade in East Asia between 1980 and 1989; Figure 6.1 compares East Asia's growth with that of the other major regions, including the same period.

The West Pacific Rim's share of world trade increased from 14.3 per cent in 1971, to 15.3 per cent in 1976, 18.2 per cent in 1981 and 22.8 per cent in 1984. Over the period from 1971 to 1984 exports from the region grew at an average annual rate of 17.9 per cent compared with world

Table 6.1   Growth of merchandise trade, 1980–89

	Merchandise trade (billions US$) 1989		Average annual growth rate (%) 1980–89		Terms of trade 1989 (1987 = 100)
	Exports	Imports	Exports	Imports	
China	52,538	59,140	11.9	11.7	104
Indonesia	21,773	16,360	2.4	−0.4	97
Philippines	7,747	10,732	1.3	0.4	107
Malaysia	25,053	22,496	9.8	3.7	97
South Korea	62,283	61,347	13.8	10.4	108
Taiwan	66,475	50,523	13.4	9.6	112
Hong Kong	28,731	72,154	6.2	11.0	100
Singapore	44,600	49,605	8.1	5.8	98
Japan	275,000	207,536	4.6	5.4	96

Source: World Bank (1991).

exports, which grew at an average rate of 14.2 per cent over the same period. Intra-regional trade provided an important element of this growth. Over the same period (1971–84) intra-regional exports grew at an average annual rate of 19.1 per cent, compared with an average annual growth rate in extra-regional exports of 17.4 per cent. By 1987 Japan, the four NICs, the Philippines and Malaysia accounted for almost 25 per cent of world trade.

These high rates of export growth in the countries of the region relative to world averages, and hence their increasing share of world trade, reflect the success of the export-led growth strategies pursued in most countries of the area. But this success in turn reflects the ability and readiness of the peoples of these countries to grasp the opportunity to undertake programmes of rapid structural adjustment, the development of new products and the exploitation of new markets.

The growth in the proportion of trade turnover accounted for by the West Pacific Rim countries as a whole is impressive and continuing. Yet some countries have done much better than others. Perhaps the most striking case is South Korea where, between 1965 and 1983, growth rates in international trade were remarkable. In international trading terms, South Korea rose from obscurity to rank fifth among all countries in the Pacific region (including the East Pacific Rim). During this period Singapore rose from fifth to fourth place; Japan remained in second place, behind the United States; and Indonesia rose from thirteenth to eighth place. On the other hand, Malaysia and the Philippines lost ground, the Philippines actually dropping out of the top ten countries in ranking during this same period.

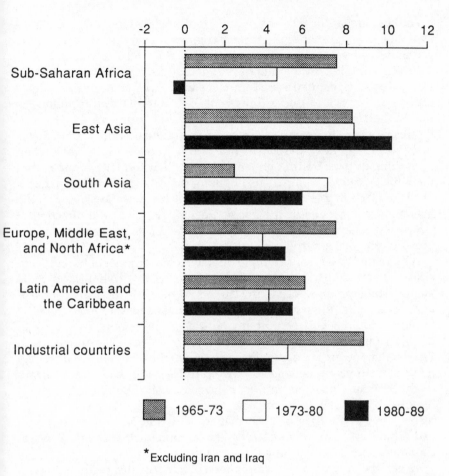

Figure 6.1 Estimated percentage annual growth in real exports, selected regions, 1965–89 (Source: World Bank data)

Direction and composition of trade

Within the Pacific Region as a whole the flow patterns of trade turnover are very simple. The North Pacific Axis, between the West Pacific Rim countries and the United States and Canada in the east is the most important, accounting for some 15 per cent of world trade. The East Pacific Axis between Canada and the United States in the north and the Latin American countries in the south accounts for a rather smaller percentage of world trade. And finally, there is the West Pacific Axis, stretching from Japan in the north down to Australasia and Oceania in the south; though relatively small at the moment, the potential for increasing international trade along this axis is enormous.

Looking first at the *direction* of trade indicated in Table 6.2, there is no evidence to suggest that the former colonial powers have been able to exert the degree of trading dominance in their former colonies as is sometimes implied in the literature. Even in the one existing colony of Hong Kong, the United Kingdom takes only 7 per cent of the colony's exports and provides under 3 per cent of its imports; and in Indonesia the Netherlands takes only 3 per cent of that country's exports and figures nowhere in its imports. It could be argued that the United States (for the Philippines) and Japan (for South Korea and Taiwan) retain something of their trading pre-eminence in their former territories. But the United States and Japan also figure strongly in the trading patterns of all countries in the West Pacific Rim simply because they are by far the major trading nations in the Pacific Region. As far as the direction of trade is concerned, there also seems to be no firm correlation with previous colonial control.

Within the West Pacific Rim the direction of international trade is very clearly dominated by Japan and the United States. Significantly, the main destination of exports is the United States while the main source of imports is Japan. But a number of other interesting issues arise from the trading directions between other countries in the region. For example, the growing trade between Taiwan and China is arousing fears in Taiwan that too great a dependence on trade with China might affect the political relationship between the two countries: it is feared that Taiwan might become a hostage, in rather the same way, though for different reasons, that Hong Kong is a hostage of China. It is partly for this reason, and in the light of 1997, that academics in Taiwan, Hong Kong and China are currently debating the feasibility of forming a 'Great China Economic Zone' of the three countries.

As for the *composition* of international trade in the region, there is some firm evidence to support the suggestion that the colonial pattern of trade is still in existence (Tables 6.3 and 6.4). Especially if we divide trade flows to distinguish between flows of crude materials (raw materials and primary products) and flows of manufactured goods, then the colonial pattern is still very much alive (Gibson, 1990). This is particularly true of the Southeast Asian sector. Despite the growth of the industrial sector in Southeast Asia, a majority of the imports into the leading trading countries in that part of the West Pacific Rim are still manufactured goods (Table 6.3), whereas a clear majority of the exports are crude raw materials (Table 6.4). In most of Southeast Asia, except Singapore, there is still evidence of the colonial pattern of trade in which countries export raw materials and primary products and import most of their manufactured goods. This contrasts sharply with Japan, where the opposite occurs: 55 per cent of its imports are crude materials and primary products and 99 per cent of its exports are manufactured goods.

Table 6.2   Direction of trade, 1989

	Main destinations of exports (%)	Main sources of imports (%)
China	Hong Kong (32) Japan (15) US (8) UK (5) Singapore (4)	Japan (29) Hong Kong (13) US (11)
Indonesia	Japan (45) US (20) Singapore (8) Netherlands (3)	Japan (30) US (14) Singapore (9)
Malaysia	Japan (23) US (20) Singapore (18) South Korea (5) UK (3)	Japan (22) US (19) Asian countries (21) EC (13)
South Korea	US (40) Japan (18) Hong Kong (5)	Japan (33) US (21) Malaysia (3)
Taiwan	US (44) EC (15) Japan (13) Hong Kong (13)	Japan (34) US (22) EC (15)
Hong Kong	US (34) China (18) FDR (7) UK (7) Japan (5)	China (31) Japan (19) Taiwan (9) US (8) South Korea (5) Singapore (4) UK (3)
Japan	US (35) South Korea (5) China (5)	US (25) China (3) South Korea (3) UK (3)

Source: World Bank (1991); World of Information (1990).

As for Hong Kong and South Korea, a clear majority of both their imports and exports are manufactured goods.

As a corollary to the increase in the share of manufactures in the total exports from countries of the West Pacific Rim, there has been a decline in the proportion of foodstuffs and agricultural raw materials in total exports. Overall, it is clear that changes in the composition of exports from countries in the region have been generally toward maufactured goods and away from foodstuffs and agricultural raw materials.

The composition of imports has similarly changed. There has been a general trend for foodstuffs to comprise a smaller share of imports, reflecting the tendency for the proportion of income spent on food to fall as incomes rise and the trend towards higher agricultural productivity and food self-sufficiency in a number of countries. Similarly, the proportion of agricultural raw materials and ores and metals in total imports has declined in most of the region, a trend consistent with increased emphasis on value-added manufacturing activity.

The rise of protectionism

The point has already been made that it was the openness of United States markets that enabled the economy of Japan and some other East Asian economies to recover so quickly after World War II. One of the repercussions of this help, however, is the huge trade deficit in the United States (Figure 6.2) and the current fears of protectionism in the United States and elsewhere. At the time of writing the Uruguay round of General Agreement on Tariffs and Trade (GATT) talks is in some difficulty over this whole matter. Over half of the US trade deficit is with countries in the West Pacific Rim. Protectionism in the United States, Canada and the EEC could lead to the countries of the West Pacific Rim drifting into retaliatory protectionism, resurgent nationalism and economic stagnation. Some authorities argue that just as open international trade has been the key to the region's prosperity, so protectionism could be the cause of its decline.

It is clear that the United States trade deficit has now become a problem for the US trading partners, and in particular for Japan, Taiwan, South Korea and Hong Kong. This is manifested in the current conflict between the United States and Japan over United States' exports, more particularly of motor cars: Japan takes 30 per cent of the United States' car market, whereas the United States takes only 0.5 per cent of Japan's car market. But two points need to be made here. One is that the Japanese market is clearly resistant to American cars primarily because they are deemed to be too large, less well-built and less environmentally acceptable than Japanese cars. The second is that the United

Table 6.3 Structure of merchandise imports, 1989 (% share of merchandise imports)

	Food	Fuels	Other primary commodities	Machinery and transport equipment	Other manufactures
China	9	3	10	31	47
Indonesia	8	8	10	38	37
Vietnam	8	23	2	37	30
Philippines	11	13	7	20	50
Malaysia	11	5	6	45	33
South Korea	6	13	17	34	30
Taiwan	7	9	13	37	34
Hong Kong	8	2	5	26	59
Singapore	7	14	5	42	33
Japan	16	21	18	14	31

Source: World Bank (1991).

Table 6.4 Structure of merchandise exports, 1989 (% share of merchandise exports)

	Fuels, minerals, metals	Other primary commodities	Machinery and transport equipment	Other manufactures	Textiles and clothing
China	11	19	7	63	25
Indonesia	47	21	1	31	9
Vietnam	12	75	2	10	5
Philippines	12	26	10	52	7
Malaysia	19	37	27	17	5
South Korea	2	5	38	55	23
Taiwan	2	6	36	57	15
Hong Kong	1	2	23	73	39
Singapore	18	9	47	26	5
Japan	1	1	65	32	2

Source: World Bank (1991).

States still believes that its former client states in the West Pacific Rim are committed to liberal trade policies whereas, as will be suggested below, they have always practised selective protectionism in a number of ways.

Protectionism, the raising of tariff barriers, the continuation of subsidies, and the failure of GATT to keep open the channels of world trade could well lead within a decade or so to the creation of separate trading

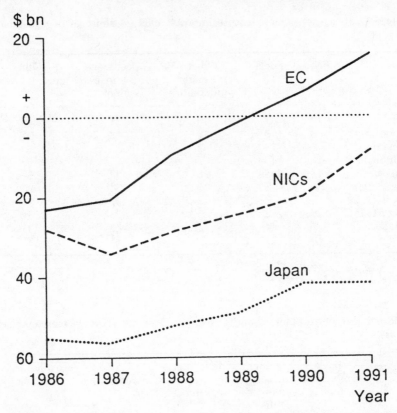

Figure 6.2 United States: trade balances with the European Community, the NICs and Japan, 1986–91 (Source: OECD; Bank of Japan; US Department of Commerce; *The Economist*, 8 June, 1991)

blocs – the Americas, Europe and East Asia – a possibility that might have more serious repercussions for Europe and the Americas than for East Asia. With over half the world's population, and containing Japan, which many believe will by then have an economy as large as and probably wealthier than that of the United States, the power of East Asia to influence what happens to the world economic order might well become critical.

It is, indeed, difficult to exaggerate the potential for international trade, both with other parts of the world and within the various countries of the West Pacific Rim. Japan is strengthening its trading and investment relationships with Australasia, the former Soviet Union and East European countries, the Americas and the countries of the EEC – where the United Kingdom, as the leading host of Japanese investment in Europe, seems set to increase its share of Japanese overseas investment after 1992.

Within the West Pacific Rim, Japan seems to be best placed to tap the almost limitless market in China, and is already beginning to dominate the trade and investment structures of most countries in the region. However, the four NICs are beginning to compete increasingly successfully in this intra-regional trade and, as indicated in Chapter 4, the Overseas Chinese in Southeast Asia have already established a close network of business links with the Chinese in Taiwan, Singapore and Hong Kong. If China's trade expands as expected, these Overseas Chinese in Southeast Asia will be well placed to take full advantage of the business opportunities open to them. Working together, the Japanese and Chinese, including the Overseas Chinese, are likely to develop the already impressive existing trading networks within the West Pacific Rim countries.

Foreign investment and technology transfer

Since the end of World War II foreign investment has been a crucial element in the economic development of most countries in the West Pacific Rim. As already emphasised, the United States invested heavily in Japan and later in many other countries in the region, and it was this investment, together with an open trade policy, that made possible the region's initial industrialisation.

In China, however, it is only since 1978 that the government has begun to encourage direct investment from abroad. Here the climate for foreigners has improved markedly over the past decade or so, indicating that foreign investors are quite prepared to invest in non-market economies if they can manufacture more competitively there, have secure access to domestic markets, and if they are allowed to repatriate reasonable profits. Athough investment flows were reduced a little after the events in Tiananmen Square in 1989, they are now back to pre-1989 levels. Either openly or covertly, many countries in the region are now trading and investing in China's economy. Thus the remarkable development in southern China's Guangdong Province has been made possible by very substantial investment, especially from Hong Kong, which provides some four-fifths of the foreign investment into the province; Taiwan provides about one-third of the foreign investment in Fukien Province; and South Korea, especially through its *chaebol* or large conglomerates, is developing similar investment links with northeast China, Liaioning and Siberia. Technology transfer is an inevitable part of this investment programme. China's Open Door policy has invited foreign participation in China's economic development, and is manifested in a major expansion of the amount and variety of technology flowing into China. Technology transfers have increased in such areas as joint ventures, licensing, equipment imports, and personnel training, including the placement of over 35,000 Chinese students in colleges and universities abroad.

China's economic strategy, like that of many of its more successful neighbours, now focuses on the infusion of foreign technology to develop exports rather than depending solely on substitution for imports. China has minimised its dependency on foreign goods, but now needs exports to pay for technology imports and other goods until indigenous technologies and production capabilities are developed.

The constraints on this policy of investment and technology transfer, however, include a unique mixture of political factors. For example, investment from Hong Kong is inevitably affected by the fact that the colony will become part of China in 1997. In spite of guarantees about Hong Kong's economic role after that date, there is naturally a good deal of caution and apprehension about exactly what will happen to investment opportunities. Moreover, Hong Kong is worried about whether China will be able to retain its most-favoured nation status; if withdrawn, this could seriously harm Hong Kong's prosperity, as the colony handles much of China's trade with the United States. Taiwan, while developing its investments in China, is anxious to avoid giving China the impression that it is too dependent on the mainland, thereby encouraging China in its claims on Taiwan as an integral part of China. Further north, similar political fears abound in South Korea. The obvious complementarity of the economies of South Korea and northeast China have encouraged trade, investment and technology transfer. But South Korea is cautious about China's complicated rules and regulations governing foreign investment. Both sides, too, have questions to ask about the kind of technology to be transferred. South Korean businesses are keen to transfer low technology and marginal industries that have been hit by rising labour costs and the rapidly appreciating *won*. China, however, is not keen to become simply a dumping ground for South Korea's offshore industries, and would prefer to see the import of more advanced technology. But South Korea is worried about the transfer to China of advanced technology, as this might increase China's competitiveness in industrial production too rapidly.

It is probably true that the major reason why enterprises move their operations offshore into other parts of the West Pacific Rim has been and remains the search for lower labour and land costs. Certainly it is for this reason that much foreign investment has taken place from Japan and the NICs, and foreign investment in Southeast Asian countries is already providing much larger shares of total investment. Taiwan's investment in Malaysia is already larger than that from Japan. But there are already several examples of wage rates rising to the extent that existing offshore operations are being forced to move on again to new locations, notably

Vietnam and China, where labour costs are still attractive. But one important disadvantage of this kind of overseas investment is that it tends to delay the eradication of low-technology industries. In Southeast Asia, in particular, there is now something of a showdown in foreign investment, leading to current account deficits: Indonesia, because of its oil exports, is now the only country with a trade surplus with Japan. As one report puts it, 'Southeast Asia's halcyon days are over' (*The Economist*, 1991b, p. 12).

The role of multinationals in investment and technology transfer in the region has not been constrained by the political and ideological objections that have often characterised their involvement in many other parts of the world. Among the NICs, Singapore has been the most energetic in attracting multinationals: there are now rather over 3,000 multinational corporations (MNCs) based in the country, and Singapore's initiative in setting up the growth triangle between Singapore, Malaysia and Indonesia, designed to attract foreign investment, is certain to increase still further the role of multinationals in trade, investment and technology transfer.

The larger companies in Japan have only quite recently begun to go truly multinational. This change has come about for reasons very similar to those which forced American companies to go multinational in the 1950s: high wages and a hard currency at home and creeping protectionism abroad. Previously, Japanese manufacturers have had their corporate headquarters at home in Japan, making them simply Japanese corporations with overseas subsidiaries. Traditional MNCs like IBM, Ford, Unilever and Nestlé have tended to operate globally with regional headquarters. Rather than simply moving some of their operations offshore to other parts of the West Pacific Rim, Japanese firms must decide whether to go truly multinational.

Conclusions

Central to the whole issue of international trade and investment in the West Pacific region is the tension that has arisen in the trading relationship with the United States, and particularly between Japan and the United States (which currently has a massive trade deficit with Japan, the world's major creditor and banking nation). Part of the tension has arisen from a misunderstanding about the nature of international trade as practised in East Asia. As indicated earlier, whereas the United States has followed liberal policies in international trade, East Asia has in practice adopted mercantilist or neo-mercantilist rather than liberal policies. Most countries in the West Pacific Rim have embraced the market, private profit, incentives and competition, but in a very different framework of cartels, subsidies, 'managed' trade, industrial policy, government intervention in the market, and the mixing of state and private interests. This contrast between openness, especially towards imports, in the United States and the mercantilism of Japan and East Asia has put the international trading system under pressure. It has been predicted that in the 1990s there will be continuing tension, resulting in a relative decline in international liberalism and a rise in international mercantilism. The success of Japan and the West Pacific Rim, based not on liberal international trade but on international mercantilism, has signalled the end of United States' hegemony in the world economy (Gourevitch, 1989, p. 15).

7

Economic strategies and government intervention

Underlying much of the discussion on the different economic strategies of the various countries of the West Pacific Rim is a tendency to imply stereotyped oppositions: communist and non-communist; command/ centrally planned and free-market economies; maximum and minimum government intervention; agriculture and industry; rural and urban; import-substitution and export-oriented industrialisation; or concentration and diversification. Certainly the region has experienced a great range of development strategies, even within the last decade, and the examination of these different strategies is useful for any assessment of the reasons lying behind the different levels of success. However, very few of these categories or oppositions are nearly as clear-cut as is often suggested; nor in most cases are they mutually exclusive or fixed categories. There is much blurring and constant change.

Nevertheless, it seems useful to begin by making a distinction between (i) the two communist, command or centrally planned economies of China and Vietnam, and (ii) the other nine countries which can broadly be termed market-oriented, largely capitalist economies. The main qualification to make, however, is that in no way does this distinction correspond to degrees of government control, interventionism or authoritarianism.

Communist economies

The centrally planned, communist economies, their economic strategies at one time determined very largely by adherence to Marxist-Leninist theoretical prescriptions, are beginning to be amended in the light of experience, both within these countries and within the communist world as a whole. The clearest and most important case is of course China, now the only large surviving communist economy.

Nevertheless, while it has remained firmly a communist country since 1949 it has experienced several changes of emphasis in economic strategy. Since 1978/79, in particular, China has followed a more pragmatic, 'open-door' policy and is aiming to create a 'socialist guided market economy'. The economy remains largely centrally controlled: its industry, for instance, is still dominated by over 100,000 state-owned enterprises, producing 77 per cent of industrial output, and some 304,000 collective (co-operative) enterprises, providing 22 per cent of industrial output. But the economy as a whole is less rigid and monolithic than formerly. The Responsibility System, linking household remuneration to output, has led to some large-scale restructuring towards a more mixed economy governed by legislation; and this has been reasonably successful, in spite of severe inflationary problems. At the same time, China's economy has been opened up to international trade, and special privileges regarding foreign trade and investment have been given to Fujian and Guangdong provinces in southern China. The five Special Economic Zones (SEZs) and fourteen ports designated as Open Coastal Cities are empowered to offer favourable incentive and investment conditions.

Vietnam's economic strategy has been based on a highly centralised communist five-year plan system. It still seems determined to maintain its political purity, even if that means not being able to introduce all the necessary economic liberalisation measures. Certainly since the end of the war in 1975 Vietnam's economy has been in a very poor state: there have been frequent shortfalls in food supplies and high rates of population growth; and a vast army of 1.3 million men has imposed severe restraints on development. Since 1986, however, the government has been following more pragmatic policies. Foreign investment is now encouraged, private enterprise is developing, and there has been a remarkable improvement in food supplies to the extent that Vietnam has become an important exporter of rice.

In discussing the communist countries of the West Pacific Rim there is a tendency in the literature to over-emphasise the limitations, constraints and ineffectiveness of their economic policies, as if they have failed completely to bring about any economic improvement at all. Later chapters, however, will reveal that there have been positive features, especially in China and, most recently, in Vietnam. While their present economic conditions and immediate future prospects do not compare with those of many other countries in the West Pacific Rim, they do generally compare favourably with those of many developing countries in other parts of the world. What seems to influence opinion about these two countries is not so much the fact that their governments follow particular policies, but that they are politically authoritarian and repressive regimes.

In the immediate future the major problem facing these two communist countries is very similar to that faced earlier by the former Soviet Union and the former communist states of Europe. One work refers to 'the dilemma of communist states attempting to reconcile the requirements of political discipline and economic liberalisation with one force – the power of the Party – ultimately unable to accommodate some of the political implications of economic reform' (Benewick and Wingrove, 1988, p. 4). The difficulty is to decide in what order and at what speed political freedom and economic liberalisation should be implemented. To date the two communist states of the West Pacific Rim seem determined to learn the lessons of what has happened further east. They seem ready to implement *perestroika*; but they are not prepared to encourage *glasnost*. They seem determined to retain political and ideological control as communist states, while allowing certain levels of economic liberalisation to take place.

This point has been developed by Prybyla (1987), who focuses on the need for institutional change. He draws a distinction between adjustments and reforms (Figure 7.1), arguing that most socialist countries have so far only shown adjustments, not reforms, producing what he calls 'birds in cages'. According to this thesis, neither leftist nor rightist adjustments will work because they ignore the need for institutional reform. Thus China has discovered that the market and private property rights, once introduced, tend to take over. Adjustments to the 'right', if carried far enough, will end up in systemic transformation ('capitalist restoration'). Real reforms of the system are excluded or inhibited by economic, political, ideological and psychological difficulties. Without proper institutional reform, which must include scrapping many existing institutions, there can be no true reform – only adjustments (Prybyla, 1987, p. 260).

Other authors also emphasise the need for institutional change. Gourevitch (1989) argues that institutions are as vital as markets, for markets cannot exist without institutions created by the state. The problem is not one of state versus markets. The problem is to find the right interaction between specific institutions and particular market situations. As the 1991 World Development Report puts it, 'history shows, above all, that economic policies and institutions are crucial' (World Bank, 1991). The principal theme of the report, indeed, is not the false dichotomy of *laissez-faire* interaction versus government intervention; rather is it the interaction between governments and markets. In other words, the report stresses the complementarity of government and markets – what it calls a 'market-friendly' approach.

Competitive markets provide the best way yet found for efficiently organising the production and distribution of goods and services; and . . . domestic and

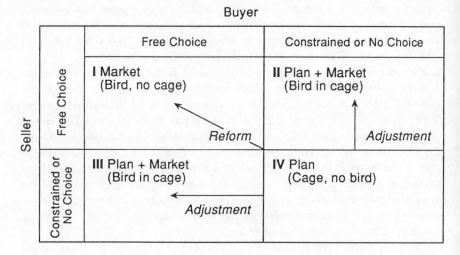

Figure 7.1 Prybyla's 'Bird in Cage' model of adjustment and reform (Source: Prybyla, 1987, p. 21)

external competition provide the incentives that unleash entrepreneurship and technological progress. But markets cannot exist in a vacuum. They require a legal and regulatory framework that only governments can provide. (World Bank, 1991, p. 1).

Market economies

One conventional explanation of the success of most West Pacific Rim countries is that their economies have grown because they got the prices right, basing their policy on the notion that all that is necessary for economic growth is to create a free market; everything else will take care of itself. According to this view, state intervention is always sub-optimal; only the free market can provide right information about real costs and provide the incentives to achieve maximum efficiency.

But, clearly, there are now elements in the economic strategies of the so-called command economies which are also market-oriented, including the open-door policy on foreign trade and investment and the encouragement of private enterprise referred to above. Increasingly, therefore, the distinction between centrally controlled economies and market economies is becoming one of degree rather than of content.

This blurring of the distinction in practice between command and market economies can be illustrated by referring to the view that the most critical difference in economic strategies among the states of the West Pacific Rim is between import-substituting industrialisation (ISI) and export-oriented industrialisation (EOI). It is clear from an analysis of the changing economic strategies in the countries of the region over the last few decades that ISI has usually been characteristic of the early stages of development and that EOI has usually been introduced later to form the basis of more recent economic growth and prosperity.

In fact, however, West Pacific Rim countries have never followed this prescription entirely: they have followed generally mixed strategies. Indeed, ISI and EOI are not mutually exclusive, and elements of ISI have remained in place everywhere. What has occurred is a shift in emphasis towards EOI, and this has commonly been forced upon governments, not because of any ideological change, but simply because ISI can only work while it is supported by a growing domestic demand. But domestic demand has often been limited either by the size of the population or, more usually, by low purchasing power, itself frequently determined by a sluggish agricultural sector. These constraints have made the logic of moving towards EOI irresistible. As White and Wade (1988) point out, for China, South Korea and Taiwan, 'they have all combined the two strategies (ISI and EOI) both sequentially and concurrently' (p. 25). Gereffi (1989) also challenges the adequacy of the distinction between ISI and EOI policies, noting that true free-market policies have not been followed in East Asia. Most governments have followed mercantilist rather than free-market strategies.

It is revealing to look at the case of Japan. During its attempt to industrialise in the last quarter of the nineteenth century, the government first followed a policy of import substitution for its cotton-spinning industry; only later was export expansion encouraged. The weaving industry also changed emphasis from ISI to EOI. Japan's light industries were built up initially under the policy of import substitution and only later, after the introduction of modern technology, did it develop the export of consumer goods. This experience contrasts sharply with what happened in many developing countries which attempted to carry out import substitution policies in the 1950s. They were unsuccessful, largely because domestic demand was not great enough for them to maintain large-scale production. Because import substitution failed, export expansion was difficult. As a result, their balance of payments position worsened because of increased imports of raw materials, production equipment and technology. But in Japan the domestic market for consumer goods was already well-developed as a result of the expansion of agriculture, commerce and traditional industries before the start of industrialisation; and exports of primary products such as copper, tea and raw silk paid for the importation of raw materials, production equipment and technology for industrialisation.

Agriculture versus industry

Relatively little is heard about agriculture in the literature on the economic development of many countries in the West Pacific Rim. There is the clear implication that the emphasis must be on industry because industrialisation is essential for economic growth. It is argued that only a country like Brunei, with a small population but great oil wealth, can expect to achieve high levels of per capita income without industrialising. The fact is that no agriculture-based economy has ever achieved per capita incomes over US$500 for any significant period (Kahn, 1979, p. 118). Industrialisation may not be the end of the road, but it is the path most countries must follow to reach higher per capita incomes.

Nevertheless, with the exception of the city-states of Hong Kong and Singapore, all countries in the region have had – in some cases they still have – significant agricultural sectors. Even as late as 1960 a number of economies in the West Pacific Rim derived a larger share of their income from agriculture than do the poorest developing countries today. And in terms of proportions of their GDP and of their work-force employed in agriculture, a number of countries have very strong agricultural sectors. Furthermore, a country like Malaysia is still largely a commodity exporting country, gaining much of its wealth from the export of agricultural commodities like palm oil and rubber (Table 1.2).

It is also clear that agriculture has in almost all cases been the basis on which successful industrialisation has been built. As already noted, this was particularly true of Japan. There is now a general realisation that all governments in the region 'must create an agricultural surplus to get their industrial sectors going'. (Lee Kuan Yew, quoted in *The Economist*, 1991b, p. 24). Rich and industrious rice farmers have been the foundation of the region's industrialisation. In Japan, South Korea, Taiwan and China, investments to make farming more productive have been accompanied by radical land reform as a necessary precondition for agricultural improvement. In the Philippines the failure to introduce serious land reform measures is certainly an important reason why the country lags behind most of its neighbours in both agricultural and industrial development.

The case of Vietnam is particularly revealing in this respect. Though the economy of this communist country is still very weak and one of the poorest in the world, it has nevertheless experienced a remarkable improvement in its agriculture, notably in rice production. Its Sixth Party Congress in 1986 proposed major reforms and initiatives. Rice output in that year was far below what it had been in 1942, and co-operative farmers in the north were producing far less rice than private farmers. Land laws were modified to guarantee farmers a ten- to fifteen-year tenure on the land they cultivated, with the expectation that land could eventually be owned and inherited. This policy change signalled the end of collectivised agriculture in the south. Moreover, since 1986, farmers have had the legal right to sell their produce on the free market after paying tax based on output. Part of this tax (10–20 per cent) goes to the state for the farmers' use of co-operatively owned machinery and for fertiliser and other necessary items. Under this new system, farmers can keep a far larger percentage of their output than was true under the former contract system. By 1989 – in less than three years – Vietnam had re-emerged as a major rice exporter, becoming the Third World's largest rice exporter.

Similar success in agriculture has been achieved in China, as well as in many other countries in the West Pacific Rim. Reforms have generally been much more successful in agriculture than in other economic sectors.

Changing strategies

It is clear that no government in the West Pacific Rim has followed a single, clear economic strategy or set of strategies for growth. Whether communist or non-communist, strategies have frequently changed in emphasis or direction. Reference has already been made to China where, since 1949, there have been quite fundamental switches in policy, most notably during the Cultural Revolution and, most recently, in the Reforms of 1978. Attention has also been directed at changes in the strategy for agricultural development in Vietnam, changes which were made possible by broader strategic changes set in train by that country's government. After unification between North and South Vietnam in 1975 Hanoi imposed upon the South the Stalinist-Maoist strategy of economic development which had until then been applied in the North. This 'Northernisation' resulted in an economic crisis for the whole country during the last years of the Second Five-Year Plan. Despite some partial reforms, the country was again plagued with a more serious economic and financial crisis at the end of the Third Five-Year Plan, particularly after the disastrous monetary reform of 1985. At the time of the Sixth Party Congress in 1986 the new leadership advocated a strategic shift in its overall economic policy under the banner of *Doi Moi* (Renovation); and it was this final change that made possible the successful changes in agricultural policy referred to earlier (Vo Nhan Tri, 1990).

The case of Brunei illustrates another kind of change in policy, from concentration to diversification. Very wealthy, but highly concentrated and almost entirely dependent on oil and oil products, Brunei's economy is vulnerable to changes in world demand and price, as well as to the inevitable threat of the exhaustion of its oil resources. The government has therefore decided to restrict oil production and to encourage diversification. This involves not only investment in a wider range of agricultural products, but also industrialisation. Another example is Indonesia, whose economy is also being diversified away from an excessive dependence on oil to include a broader and more efficient agricultural base as well as what is now a rapidly burgeoning industrial sector.

Taiwan, too, has periodically altered its growth strategies. In Taiwan the First Four-Year Plan was introduced in 1953, adopting ISI policies, partly because Taiwan's export market for food to Japan had been cut off. By 1959, however, the emphasis had shifted to EOI and by the early 1970s, when Taiwan became an NIC, a change in strategy concerning the balance of labour and capital was forced on the government by a labour shortage. Capital-intensive industries were now encouraged. In spite of set-backs like the oil crisis of 1973–74, Taiwan has continued its success by adjusting to higher energy costs and by encouraging the growth of a petrochemical industry. In the 1980s there were high domestic labour costs and severe international trade competition, difficulties to which Taiwan responded by developing a computer and electronics industry (Wade, 1990).

Similar changes in strategy seem to have occurred in all the countries of the West Pacific Rim. But this is not to suggest that all economies have gone or should go through the same changes of direction and emphasis. It is true that many countries seem to have followed or are following a broadly similar sequence: agricultural progress, a short period of emphasis on ISI, a gradual shift of emphasis to EOI, and a policy based on comparative advantage in international trade. But the detailed effects of these changes and the degree to which they have been successful or permanent seem to have depended and will continue to depend on the nature, if not the extent, of government intervention and on a unique mix of circumstances, opportunities and constraints in the individual economies.

Government intervention

It is sometimes implied that government intervention is only characteristic of command, communist countries, and that free-market economies are characterised by little or no government intervention. According to some writers on East Asia, governments began in the 1960s to allow market forces to operate. Progressive liberalisation and opening up of the economies to outside markets have subsquently ensured success. Success in Japan, for instance, has been due to private individuals and entrepreneurs responding to the opportunities provided in free markets. The only useful role governments have performed is to provide supportive environments in which entrepreneurs are enabled to perform their functions efficiently. Friedman's view on this is quite categorical. In arguing for reducing the role of government he claims that 'Malaysia, Singapore, Korea, Taiwan, Hong Kong and Japan – relying extensively on private markets – are thriving . . . By contrast India, Indonesia and Communist China, all relying heavily on central planning, have experienced economic stagnation' (quoted in Wade, 1990, p. 22).

But even the most preliminary study of the countries of the West Pacific Rim shows this to be quite inaccurate. In all eleven countries of the region, even including the British colony of Hong Kong, governments have intervened quite heavily. In the Philippines it is believed that the government, whether local or national, is able, given the political will, to plan for and to effect economic development.

> With careful planning and by the considered and determined execution of these plans, it should be possible to construct a successful economy which to some degree conforms with the prevailing view or ideology of what economic development should be. Implicit in this belief is the assumption that, without the execution of these plans, economic development will be limited, unbalanced and detrimental to the majority of the population. (Hodder, 1991, p. 105)

In Taiwan and South Korea the governments intervened strongly in the 1950s when ISI strategies were generally adopted throughout much of East Asia. And it has been pointed out that

> the view of Singapore as the archetypal *laissez faire* economy could not be further from the truth. However, this is not to say that it is not a capitalist economy, but rather that its success has not been built upon the operation of the free market. The government has been highly interventionist and has very carefully, and very successfully, for much of the time, stage-managed Singapore's development. (Rigg, 1991, p. 196).

Clearly it is false to make any precise distinction between planned or command economies and those that profess to be market economies; indeed, the consequences of dichotomising choices between planning and markets in this way are serious both for theoretical analyses and for practical policy-making. In all the economies discussed here the strategies are now mixed, whether they are described as command or free-market economies. This is logical in the sense that the present trend in thinking is to emphasise the mixed nature of economic development strategies. Elster and Moene (1989) argue for middle paths; and even in the NICs, where the liberal free-marketeers seem to have clear evidence to support their views, government intervention has in practice gone far beyond what would be approved of by most economic liberals.

In his influential book *Governing the Market*, Wade (1990) considers the polarised views on the role of government in the economic development of East Asia. At one end of the ideological spectrum are the neo-Marxists and dependency theorists who emphasise the need for government and socialist state institutions to create the necessary preconditions for successful and socially equitable economic development. At the other end there are those who adopt a neo-classical analysis and emphasise the need for relatively free and open market forces to operate. Basing much of his material on the Taiwan case, Wade documents the critical role played by the state. What is important, he argues, is to realise that neither of these polarised positions is valid. What needs attention in the analysis is the *way* in which governments allocated decisions. 'The government did not simply control markets; it also offered periodically updated visions of the appropriate industrial and trade profile of the economy and gave a directional thrust to private sector choices in line with these visions' (Wade, 1990, p. 4). In other words, East Asian industrialisation has depended critically neither on *laissez-faire* development policies nor on central government control, but on selective, guided but extensive government intervention.

It seems that it is not the degree of government intervention, but its nature and motivation that matters. In a comparison between the East Asian and Latin American economies, it has been noted that:

> whereas the liberal analyst says that Latin America governments intervened too much in the market, and that is why their countries performed less well than East Asia, we find that state intervention in the latter has been both stronger and more selective than in the former, not only at the national boundary but also in key parts of domestic industry. The aim of that intervention is to build up national powers of production, reduce national vulnerabilities and, to some extent, to minimise the socially disruptive costs of market adjustment. (White and Wade, 1988, pp. 9–10)

In the Philippines, the reaction against an indiscriminately interventionist government is clear: there are leading politicians who are unequivocal in their opinion that:

> the role of government should be to facilitate the creation of wealth – not to redistribute wealth, nor to try to mould or redirect the forces which create it. They point out that the Filipinos are cynical of any government which believes it can impose its ideals and which tries to chart a course to which all will adhere. (Hodder, 1991, p. 125).

Government intervention has been and remains a potent factor in the development of all countries in the West Pacific Rim. It has been argued that the modern notion of 'development' rests on a concept of the state as the main generator of socio-economic progress. Development in the region is 'late development', in other words, it is very different from early industrialisation of the kind experienced in Western Europe, for instance. Such late development is less spontaneous. The state plays the role of historical animateur. 'The ideology of "developmentalism" and the idea of the interventionist state are thus inseparable' (White and Wade, 1988, pp. 1–2).

The next three chapters are centred on individual regions or countries. They are not designed to give potted summaries of the geography, history or political economy of the countries, but to illustrate, however briefly and selectively, some of the points and arguments presented in the previous chapters. They also demonstrate that while it is possible to make a number of general statements and to suggest several ideas, these need constantly to be re-examined in the light of each country's experience and 'cultural signature'.

8

Japan and the four NICs

Japan

Japan requires first and separate consideration here, not only because it is one of the world's major developed industrial economies but also because it is frequently cited as a 'model' for other economies in the region. Economically it lies well ahead of the four NICs (Singapore, Hong Kong, Taiwan and South Korea) and is immensely more wealthy and powerful than any of the other 'nearly industrialised' or still-developing countries in the area. The Japanese income per capita is higher than the American, and during the late 1980s Japan emerged as the world's banker and major creditor nation. Ten of the world's largest banks are Japanese, as are fifteen of the world's twenty largest corporations (Nestor, 1990, p. ix). Japan, indeed, is the country which 'shapes and shares the dynamism of the entire Pacific world, both east and west' (Inoguchi, 1989, p. 87). It has been a member of OECD since 1965.

It is of course important to emphasise that Japan embarked upon its modern industrial development some eighty years earlier than any of the other countries in the West Pacific Rim. Therefore in answering the question of why Japan has been so successful in its economic growth, we need to take a rather more retrospective look at development in Japan than elsewhere in the region.

An important advantage possessed by Japan when it embarked upon its modern phase of economic development in the 1880s was that its pre-modern phase began during the closing stages of the Tokugawa period, when western science and technology began to be introduced. By the time of the Meiji Restoration of 1868 and the opening up of Japan, important irrigation works were already in place, and there was a reasonable network of roads, railways, postal and telegraph services and a coastal shipping network. Agricultural technology was sufficiently advanced to enable agriculture to meet the increased demand for food and capital accumulation created by industrialisation. Rural industries had also progressed as a basis for modern industry and, with the growth of industry, the domestic market grew. Education was widely diffused, literacy levels were high and an efficient political system was established (Minami, 1986, pp. 423–4).

Other advantages possessed by Japan – advantages which seem to have heavily outweighed the country's lack of natural resources for industry – include the fact that Japan has only a small land area to control, together with a climate giving abundant rainfall and widely varying seasonal temperatures – a fact which Minami, in a remarkable display of climatic determinism, suggests is 'a source of stimulation and refreshment for the people, and could be said to account partially for their energetic nature' (1986, p. 4). The population is relatively homo-geneous, speaking a single language which has facilitated the exchange of information and a sense of unity. Population densities are high in the occupied plains and, according to Minami, this means that 'contacts are very frequent, so that an efficient communication network has devel-oped naturally. This has contributed greatly to the people's ability to come to a consensus of opinion, and also to the diffusion of technology in agriculture' (1986, p. 4).

Moreover, the abolition of the old social class of the *samurai*, farmers, artisans and merchants after the Meiji Restoration meant that there were no rigid class barriers to development; this was improved still further after 1945 by the occupying powers who attempted to introduce more democracy through agricultural land reforms, the break-up of the *zai-batsu* (the giant business conglomerates) and the formation of labour unions. In 1946 Japan was given a new democratic constitution with universal adult suffrage and a constitutionally elected government. Moreover, the emperor was forbidden to engage in politics. All these acts created a high degree of equality, both socially and economically, to the extent that most Japanese today think of themselves as middle class.

There is also no seriously constraining religion. Though there are several religions – Shinto, Buddhism and Christianity – none has a limiting influence on everyday life. Traditional and modern elements are inextricably mixed in Japanese society. Influenced by Confucian teach-ing in the Tokugawa period from 1603 to 1867, the Japanese people regard the interests of the group as being greater than those of the individual. Thus they demonstrate a loyalty to the extended family, the company and the country and have a belief in the virtues of saving rather than consuming, and it is this last fact which is said to contribute to the high level of domestic savings. Several authors have commented on what one has called the Japanese 'legendary, almost Calvinistic propensity . . . for saving money, for buying with cash, for scorning credit' (Winchester, 1991, p. 279).

In trying to understand Japan's success today, however, it is easy to focus on these various advantages and to suggest that they add up to a reasonable explanation of that success. In fact, however, the above paragraphs represent the kind of physical, historical and cultural determinism already rejected in earlier chapters. It needs to be emphasised that Japan's remarkably successful industrial economy has been created in a country with virtually no natural resources, and that its economy lay in ruins at the end of the Second World War, only forty-five years ago. Ironically, it was perhaps these two factors that made Japan's dramatic economic growth possible. Her lack of natural resources has always forced the country to find ways of developing its human resources and to adopt strategies that involved importing raw materials and manufacturing exports to pay for those imports. Furthermore, it could be argued that it was the very weakness of Japan's economy in 1945 that determined the occupying Americans' build up of Japan's economic strength as a means of halting the eastward spread of communism. Investment in Japan by the United States was heavy and, as indicated in the previous chapter, open and preferential treatment was given to Japanese exports in United States' markets.

Japan is central to our discussion for many reasons. One is that the government has certainly played an important role in setting and facilitating the industrial strategies followed since the 1950s. During and since the 1970s industry has been advised, co-ordinated and given 'administrative guidance' by the 'twin juggernauts of the nation's miracle-building, the Ministry of Industrial Trade and Industry and the Ministry of Finance' (Winchester, 1991, p. 43). Japan, the so-called bastion of the free market, has in fact always been ready to take advice from its government, to the extent that it has been called 'a centrally-planned free-market economy'. Japan is also central to our discussion because of the way in which it has diffused economic benefits to other parts of the region. For example, Japanese electronics firms first moved their operations to Taiwan and Korea when wage and other costs were lower than in Japan; but as soon as costs rose to uncompetitive levels the Japanese moved their operations still further away to the Philippines and China (Inoguchi, 1989, p. 48). Another distinguishing feature of Japan's modern economic growth has been its application of modern science and technology to all fields of production. Yet the Japanese have always been fully aware that its expenditure on social overhead capital and education must be sufficient to enable industry to absorb this modern technology. Moreover, the Japanese were from the outset prepared to apply capital-intensive, borrowed and indigenous technology to agriculture as well as to industry.

Japan has been much criticised for its soaring trade surpluses (Figure 6.2), though its import-export trade (Tables 6.3 and 6.4) has recently provided a much smaller percentage of national product than in West

European countries. Most recently, too, consumer demand has been growing to the exent that its economy is shifting away from its traditional dependence on exports, resulting in a reduction in its trade surplus. 'The consumer market, always strong in Japan, has been transformed into one thirsting for imports and for ways in which buyers can express individuality rather than conformity' (*The Economist*, 1991c, p. 68). This trend has important implications, not only for Japan's trade surplus, but also for the dynamics of those cultural factors to which many writers attach so much importance in their explanations of Japan's economic success.

The one area in which Japan has a difficult problem opening up to imports is rice – at first sight remarkable for a country which is the largest importer of food in the industrialised world. For internal political reasons, however, rice in Japan is heavily subsidised by the government in order to retain the support of the three million politically powerful rice farmers, even though they are mainly only part-time farmers cultivating small plots averaging three acres each. Yet the Japanese still have to pay very high prices for their staple food. Rice costs six times more to produce in Japan than elsewhere and, even with the subsidies to soften the blow to consumers, the retail price of rice is almost three times the price of rice in the United States or Europe. This is a good example of political interests overriding simple economic imperatives, even in a country where economic efficiency is apparently of paramount importance.

Japan has many other problems, notably its high social expenditure on education, pensions and sickness benefits, especially in the context of its rapidly ageing population – an inevitable result of its successful population control policies. It also has a serious group of problems arising from excessive urbanisation and associated environmental pollution. Nevertheless, it is to Japan that the other countries of the West Pacific Rim look, for an example of and stimulus for what they hope to achieve themselves, and as the major investor in their own economies.

The four NICs

The previous chapters contain frequent reference to the four NICs, known variously as the 'Four Little Tigers' or 'Four Little Dragons', though in Japan they have been awarded the accolade 'Four Little Japans'. These four countries – Singapore, Hong Kong, Taiwan and South Korea – have achieved remarkable economic successes in the past two decades. By almost any criteria, they have dynamic prosperous economies and almost uniquely high rates of growth, and their per capita incomes are now amongst the highest in the world. Even though one of them, Singapore, lies much further south than the others in the very centre of Southeast Asia they are often treated together. Yet these four countries differ in so many ways: in their governments, strategies, developmental paths and future economic prospects.

Singapore

Singapore's success has made it a standard-bearer for those who advocate an outward-looking export-oriented strategy. Neo-classical economists have claimed that Singapore's success is the product of the 'free market' and use it to argue the case against state intervention. However, the frequent changes in economic policy over the past twenty-five years suggest that the government of Singapore has been far from non-interventionist or *laissez-faire*. The government has been flexible, firm and ready to respond to and anticipate changing circumstances and is indeed determined to intervene whenever necessary. It is probable that without state intervention in the labour market and in the social reproduction processes, Singapore's spectacular economic success might not have occurred.

A former British colony, the island city-state of Singapore achieved independence in 1959, joined the Federation of Malaysia in 1963, but seceded in 1965. At independence its economy was in very poor condition. Population growth was 4.4 per cent a year, unemployment stood at between 10 and 15 per cent, poverty and industrial unrest were widespread. There was political turmoil, a flight of foreign and local capital, very little manufacturing base to speak of, and a decaying infrastructure. Initially, the government followed a strategy of ISI, which seemed logical when it was part of the larger market of the Federation of Malaysia. But on leaving the Federation, Singapore soon realised that its small size and lack of natural resources indicated the need for a shift from ISI to EOI. This decision was further reinforced in 1967 when the British Government announced its intention of withdrawing its military bases, thereby removing economic activity accounting for 23 to 24 per cent of GNP. Because of lack of domestic entrepreneurs, technology and capital, the Singapore Government recognised that this reorientation could not be based upon domestic resources. Instead, they began to encourage foreign investment and the involvement of foreign multinational corporations and companies in Singapore. The economy was liberalised and in a series of measures the Government was successful in attracting a massive influx of foreign investment.

In general, this strategy of MNC-led, export-oriented industrialisation was spectacularly successful. But by the end of the 1970s it was decided to embark on a second bold restructuring of the economy, involving a forced shift from labour-intensive to skill-intensive and capital-intensive industries. Enterprises were forced to upgrade, largely by increasing wage costs above productivity levels. Wages rose by 42 to 46 per cent between 1979 and 1981, skills and training were tackled, the use of cheap immigrant labour was restricted as far as possible, controls on the immigration of skilled professionals were eased, and there was more investment in research and development. This 'high-cost' policy was designed to provide industrial growth, not through internationalisation, but through enforced automation and product diversification, moving into capital-intensive and labour-saving industries.

In the mid-1980s Singapore's economy went into recession and the government responded by once again changing its economic strategy. The high-wage policy was reversed, thereby reducing manufacturers' costs; the emphasis was shifted away from manufacturing and into modern services, in which Singapore is felt to have significant comparative advantage; and the economy was further liberalised by reducing – at least ostensibly – the role of government in economic management.

The *dirigiste* nature of the government's role in Singapore's economic growth is often justified, not only by its obvious success, but also by the constraints within which it has had to operate. Singapore occupies a small island of only 627 square kilometres, is completely without natural resources, and, with a population of 2.7 million, is one of the most densely populated countries in the world. With the exception of eggs, poultry and pork, it has to import all its food, most of its water, and all its industrial raw materials.

Furthermore, Singapore's population is far from homogeneous, containing as it does a number of distinct cultures associated with the Chinese (76 per cent), Malays (15 per cent) and Indians (6 per cent). And, as noted in Chapter 4, even the Chinese comprise several distinctive dialect groups, notably the Hokkiens, Cantonese, Teochews, Hakkas and Hainanese (Cheng, 1985), with what are in effect mutually unintelligible languages. Though Mandarin is strongly encouraged, English remains the lingua franca for a large part of the population. There are therefore hidden ethnic tensions within the Singapore society, tensions which the government is well aware of and has so far been able to control. The government has restricted the percentage of Malays in its Housing Development Board flats to 25 per cent and in various ways is trying to integrate the non-Chinese more fully; but the ethnic tensions remain to threaten the political stability and so the economic prosperity of the state. Aware not only of this problem but also of the dangers to the state of material success – the 'chaotic individualism' of the West – the Singapore government is attempting to create a new sense of national identity and unity through what it calls a 'communitarian' system of belief. Predictably, however, this is essentially Confucianism in which the values of loyalty and obedience are prized above all.

Nevertheless, the human resource base of Singapore is in general of very high quality: energetic, ambitious and well-educated. In 1987, 15 per cent of all those entering the labour market had degrees; and standards of education, training and health, as well as living standards, are very significantly higher than in any of Singapore's neighbours. The main labour problem in Singapore is the shortage of unskilled, as well as skilled and professional labour. This is now a very serious matter: the present labour force of some 1.3 million shows little sign of increasing. Singapore still has to depend, however reluctantly, on immigrant labour from Malaysia, Sri Lanka and Thailand. Furthermore, it is suffering from the emigration of some 3,000–4,000 members of the professional class every year, most of the emigrants going to Australia and Canada. Singapore is now trying to coax migrants from Hong Kong in the period up to 1997. The labour problem is also being compounded in the long term by the very low population growth rate. The rate of less than 1 per cent resulted to some extent from the Singapore government's earlier 'one-child-a-family' policy, supported by all kinds of sanctions and incentives; but it is generally agreed that it was economic prosperity and high levels of education that were the main factors resulting in the drop in birth rates. Population control in Singapore has always been viewed as an adjunct to rather than as a determinant of economic growth. The government is now encouraging larger families, especially for the better-educated citizens of Singapore, in what has been criticised as a form of genetic engineering.

Singapore's economic future would seem to be both bright and assured. With its current real rate of economic growth at 9 per cent, unemployment at 3.3 per cent, and inflation at 1.5 per cent, Singapore's development has outstripped those of the other three 'Tigers' as well as of most developed countries. The structure of the economy appears to be sound. The government has shown itself able to respond quickly and effectively to global economic troughs by, for instance, encouraging construction booms and a most succesful tourist industry, which brings some 3.5 million visitors into the country each year.

The government is well aware of the danger of jeopardising this remarkable economic success by loosening its political control too far or too soon. There is a good deal of control over public and private behaviour, there is still no completely free press, and the circulation of certain journals is controlled or banned even though this makes it more difficult for Singapore to achieve its aim of becoming a global 'infor-mation centre'. While in theory a multi-party democracy, Singapore is still dominated by one party, the People's Action Party (PAP), and economic success seems to guarantee the continuation of this dominance by what has been called the 'harsh paternalism' of the present government's style.

As the most successful economy in Southeast Asia, Singapore has played an important role in the work of the Association of Southeast Asian Nations (ASEAN). Its relations with the other ASEAN states, as well as with China and Taiwan, are now more positive than they have been for some time, and the strategic importance of Singapore has been given a new boost by the United States' decision to use the island's naval facilities. Of even more importance, Singapore is developing its economic links with Malaysia and Indonesia in its new 'growth triangle' policy. This involves an area within a fifty-kilometre radius of Singapore, reaching into Johore, in southern Malaysia, and into several islands off the coast of Sumatra in Indonesia. The idea – stimulated partly by the shortage and high cost of labour in Singapore – is that investors will use Singapore as their high-technology base to design, market and distribute products made in low-cost factories in these neighbouring territories. This has been taking place in Johore for some time, but is now being extended into Indonesia, where the Batam Industrial Corporation (40 per cent Singapore-owned and 60 per cent Indonesian-owned) seems to have a bright future.

Looking to its future, Singapore's problem is that it is the most vulnerable of all the NICs to a decline in world trade. It is the most trade-dependent of the four 'Tigers' and 25 per cent of its exports go to the United States (Table 6.2). It is also vulnerable in that while it enjoys heavy inward investment (with over 3,000 MNC offices on the island), its own activities in the MNC field are small. On the other hand, Singapore is less vulnerable to protectionist retaliation than Taiwan and South Korea because of its liberal trade policy and its few restrictions on quotas or on export and import licences. What Singapore needs in the increasingly competitive market is to secure a larger market share in Japan, Southeast Asia, the countries of the former communist bloc, China and the EEC. All this will have been helped by Singapore shifting its entrepôt functions from exporting Southeast Asian goods to being a gateway for goods entering the region. It will also be helped if Singapore achieves its aim of replacing Hong Kong as Southeast Asia's financial centre after 1997.

Hong Kong

A British Crown Colony until 1997, when it reverts to Chinese control, Hong Kong is similar to Singapore in some ways. It is a small country of only 1,068 square kilometres, comprising the island of Hong Kong; Kowloon and the New Territories on the Chinese mainland; and over

200 other islands. It also has no significant natural resources apart from its fine natural harbour, and it contains a very large population of some 6 million, living at densities even higher than those in Singapore. But in Hong Kong the population is much more homogeneous, being 98 per cent Chinese and without Singapore's dialect group problem: most of the Hong Kong Chinese are Cantonese refugees who fled into Hong Kong, usually without any assets at all, from Guangdong Province in China during the upheaval of the Communist takeover in and around 1949.

Hong Kong's economic success has been remarkable. It has one of the highest per capita incomes in the world, enjoys annual growth rates of about 10 per cent, has an unemployment rate of only 2.2 per cent, a rate of population growth almost as low as that of Japan, and is one of the world's major financial centres. Hong Kong is also the world's largest container port and is among the top twelve trading countries in the world; it is also the largest exporter of toys, clothing and watches. And all this has been achieved within a generation.

This success in what is in effect a largely barren territory has been achieved by a colonial government whose rule has been remarkably non-interventionist. Compared with the government control exerted in Singapore, Hong Kong has developed as a *laissez-faire* economy. The government has followed a consistent policy of free trade, providing a liberal economic environment of freeport status, low taxes, few exchange controls, and an efficient financial and business sector. Nevertheless, the government has by no means been entirely non-interventionist. It has been pointed out, for instance, that 'the free-market paradise of Hong Kong operates the second largest public housing system of the capitalist world' (Castells *et al.*, 1990, p. 1).

Yet while this environment has encouraged and facilitated economic activity, the real engine of growth in Hong Kong has been the entrepreneurial spirit fostered by the desire of its largely immigrant population to make money, and by the readiness of so many of its small businessmen to engage in risk-taking. However one attempts to explain it, there is undeniably a powerful work ethic among the people of Hong Kong; and the most notable feature of Hong Kong's manufacturing sector is the large number (over 50,000) of small manufacturing enterprises with ten or fewer employees. Moreover, the population is highly mobile and adaptable.

Although there is a significant heavy industrial sector of steel products, shipbuilding and repairing, and chemicals, the manufacturing sector is dominated by textiles and clothing. In this Hong Kong had the benefit at an early stage of receiving immigrants from China. As Deyo notes, refugees to Hong Kong from the mainland included a significant segment of the Shanghai capitalist class and a huge supply of politically unorganised labour, and they brought with them technical know-how, skills and even machinery (1987, pp. 107). The Hong Kong textile and clothing industry is now highly sophisticated, computerised and fashion conscious. The quality is very high, even by Japanese, European and American standards, and the industry's products compete successfully throughout the world. The electronics sector is the second largest sector and has much more overseas involvement, contains larger enterprises (but fewer in number), but suffers from lack of investment into research and development. As for plastics, watches and clocks, these provide the classic example of Hong Kong's quite exceptional ability to manufacture quickly and cheaply in response to shifts in the market. Finally, the non-manufacturing service sector includes not only finance, but also a thriving tourist industry, which attracts over 4.5 million visitors a year. As for agriculture, it covers only 9 per cent of this rocky territory and, even with fishing, accounts for only 0.4 per cent of GDP. The colony is therefore dependent on imported food.

In spite of its large population, Hong Kong does have a labour shortage problem, recently estimated to be over 100,000. This has encouraged wages to rise and has produced calls for more immigration into the colony. Alternatively, manufacturers have developed a large-scale shift of their operations across the border into China, either into the nearby Shenzhen Economic Zone or deep into Guangdong Province. This is now a very important aspect of Hong Kong's industrial structure, and it has been calculated that there are now more Chinese across the border (over 2 million) working for Hong Kong manufacturers than in the entire manufacturing sector in Hong Kong itself.

But easily the main problem faced by the economy in Hong Kong is political: the uncertain nature of its future after 1997, when it reverts to China's control. It is true that the colony has been promised autonomy as a Special Administrative Region (SAR) within China after 1997, and that its capitalist system is to continue for the first fifty years. But, especially since the events in Tiananmen Square in June 1989, Hong Kong fears that it cannot trust China over this arrangement. Furthermore, as already noted, Hong Kong's economy is now integrated with and dependent upon that of Guangdong. The colony is already fearing a decline in its trade, dependent as it has recently become on fulfilling once again its traditional entrepôt role for China; and it is now experiencing a flight of some of its most valuable entrepreneurs, managers and capital – a trend that is likely to increase as 1997 draws closer and as more countries, including Singapore, agree to accept quotas of immigrants from Hong Kong. Uncertainty is also affecting stability in the financial markets, epitomised in the minds of Hong Kong citizens by the move to London of the Hong Kong and Shanghai Bank's headquarters.

For all these reasons, most commentators fear that the remarkable economic success achieved by Hong Kong will not last and that its future is both bleak and uncertain, not because of faulty economic policies, but because of unavoidable historical and political imperatives. On the other hand, it is not impossible that this scenario is too pessimistic. After all, Hong Kong is of incalculable practical value to China as it attempts to enter more fully into the world economy and encourages international trade with its open-door policy. Hong Kong already exports 60 per cent more than the whole of China and gives China access to an important international financial and trading centre. Logic and China's self-interest would seem to suggest at least the possibility that the direst predictions over Hong Kong's future after 1997 are not necessarily valid. Moreover, as the next chapter will show, the economic 'miracle' now taking place in Guangdong, and its integration with Hong Kong, could in fact operate to Hong Kong's long-term advantage.

South Korea

In spite of many constraints, South Korea has shown perhaps the greatest economic improvement of all the NICs over the past twenty-five years. While it is not the richest of the NICs, it has shown startling rates of growth, averaging 7 per cent a year, giving it the reputation of a 'miracle economy'.

By far the largest of the NICs, South Korea (just over 99,000 square kilometres) is a peninsular country, though it is in some ways as isolated as the other three. It comprises the southern half of the Korean peninsula, and the border with its northern neighbour – the communist state of North Korea – is in effect a cease-fire line between two parts of Korea which are technically still at war. It is a mountainous country, lying very much further north then the other NICs; indeed, it has strongly contrasting seasons, with very cold dry winters and hot humid summers. In spite of its size, its natural resource base is very limited: it is North Korea which has the major natural resources, though South Korea has some coal and iron, and a few other minerals. But the country still has to import most of its raw materials, including all its oil, for industry.

Part of the explanation for South Korea's economic growth lies in the country's human resources. Its population of over 42 million is now growing at an annual rate of only 1.2 per cent, a dramatic decrease from the 2.9 per cent of 1961. The density of population is high and the country contains several very large urban centres, notably Seoul (10 million) and Pusan (3.5 million). The population is also very homogeneous, being almost entirely Korean and speaking its distinctive language. Even in the Far East context the Koreans are noted for being exceptionally hard workers, well-educated, and even more thrusting, individualistic and tough than their Japanese or Chinese neighbours.

Democratic government was introduced in 1988, but the government remains essentially authoritarian, though it has very recently begun to liberalise. Yet its economic policies have long been pragmatic, expressed in a *dirigiste* form of EOI. Until recently, the government favoured large, already successful companies able to earn foreign exchange, giving support to the well-known, family-run *chaebol* – which are really only self-conscious imitations of Japan's *zaibatsu* – such as Hyundai, Samsung, Daewoo and Lucky-Goldstar. Most recently, however, the government has begun to encourage smaller and medium-sized companies, though the top ten conglomerates still account for about 50 per cent of South Korea's exports.

Another characteristic of government policy has been its emphasis on heavy industry, but here, too, the emphasis is beginning to shift to light industry. South Korea, however, still remains an important supplier of steel goods to Japan and China, has a major shipbuilding industry, and is well-known for supplying heavy contract work overseas, especially in the developing world.

Unlike the city-states of Singapore and Hong Kong, South Korea has a significant agricultural sector, which employs 20.8 per cent of the workforce and contributes 12.3 per cent of GDP. Agricultural production is dominated by smallholdings, and although the government subsidises it by import controls and by a heavy tariff on imported consumer foods, there has been a clear shift from agriculture to industry in the country's economic structure.

In spite of the large population and a low unemployment rate of only 3.1 per cent, Korea has had a series of labour problems, many of them deriving from the very characteristics of the Korean people that have helped the economy to become so successful. As many commentators have noted, South Korea's economic miracle could so easily be destroyed by endemic labour unrest. Not only has there been serious unrest over wage rates, but certain elements of the working population, possibly politically motivated, have exhibited the kind of xenophobia that is almost unknown in the other NICs. This xenophobia has usually been directed at the presence of the US army, operating under the control of the United Nations; against foreign-controlled MNCs; and against some of Korea's largest companies and giant *chaebol*. The government has tried to deal with this unrest by imposing an economic package of wage and price controls, but the *chaebol*, in particular, remain a focal point for political unrest. The sales of the affiliate companies and the turnover of the top thirty *chaebol* in 1990 was equivalent to 76 per cent of the country's GNP.

But there are economic as well as political reasons why the government is now trying to reform the *chaebol* in South Korea. These giant conglomerates generally served the South Korean economy well during the 1960s and 1970s when, supported by the government with cheap credit, import barriers and political favours, they led the country's rapid industrialisation. But they are now proving to be inflexible, and the need for smaller units of organisation is obvious. Politicians seem to be finding it difficult to make the necessary changes; but this is perhaps not surprising, considering that most politicians depend on the *chaebol* for financial support.

The higher wages which now characterise the South Korean economy – though they are still the lowest of the NICs – mean that some Korean manufacturers are moving their operations to lower-cost locations in China and Southeast Asia, a movement which is occurring in all the other NICs and is developing as an important linkage factor in the economies of the West Pacific Rim. South Korea also faces the problem of access to a wider range of markets, trying to control imports in order to protect domestic agriculture and industry, and thereby coming into conflict with the very international community on which an export-based economy like South Korea must rely. Similarly, South Korea needs to open up its financial markets internationally and to encourage a freer flow of foreign capital into the country.

The United States is keen not to antagonise South Korea, in spite of charges against the South Korean government over dumping and protectionism, because of South Korea's importance in the Pacific balance of power, especially in the event of Korean reunification. South Korea is also keen to keep the support of the international community: it is aware of the competition coming from its former communist trading partners in the former Soviet Union and in Eastern Europe, as well as of the dangers of a 'Fortress Europe' after 1992. South Korea has already offered the former Soviet Union $3 billion in tied aid in the hope that South Korean firms will be able to seize its market share before the Japanese.

Economists seem optimistic that higher technology manufacture in electronic goods, cars and steel will enable South Korea's economy to flourish, even though the country's currency, the *won*, has recently been appreciating dangerously. But politically it is clear that South Korea's democracy of only three years or so is very fragile and that internationally there are many pressures: against United States forces in the country, for reunification with North Korea, and from the insecure border with its northern neighbour. Logically, and in terms of propinquity, South Korea's natural trading partners, other than North Korea, are Japan and China. Developing these contacts would also help lessen South Korea's dependence on the United States and the West. The complementary nature of the economies of South Korea and China is obvious. South Korea exports textiles, electronics, footwear and machinery to China, and imports raw materials and basic commodities, including raw materials, needed for her industries. But the South Korean government remains cautious, particularly in such fundamental areas as technology transfer – something which China needs desperately, but which, as noted in the previous chapter, South Korea is reluctant to engage in too enthusiastically for fear of competition on world markets from China's industries. However, such is South Korea's present success that the government is now guiding the economy towards becoming less trade dependent. Domestic consumption, as in Japan, is fast becoming the main engine of growth.

Taiwan

Taiwan, or the Republic of China as it is formally known, had its origins as a separate 'state' in the Communist takeover in China and the consequent flight of the Nationalist government and its supporters to what was then known as the island of Formosa. Underlying the economic success story of Taiwan is its government's determination to show that its economic policies are more successful than those of the centrally planned and controlled Communist state of China. While Taiwan's economy is based on a free-market, capitalist system, its planning is designed and carried out by government. Certainly this success has been dramatic, resulting in annual per capita GDP growth rates of 7 per cent over many years, and between 1984 and 1989 the per capita income doubled. Taiwan also has one of the lowest income disparities in the world (Figure 1.1). But only recently, since 1988, has there been an improvement in relations between the two countries. Indeed, it is now possible to envisage closer economic ties between Taiwan and China, even if these do not fully match up to Taiwan's desire to see a kind of confederation which would result in 'one China, two governments'.

A large island of 36,000 square kilometres, Taiwan is mountainous and forested in the east, but in the west a coastal plain faces China across the narrow Formosa Straits. Like the other three NICs, Taiwan has few natural resources apart from some coal, marble and dolomite, and has to import most of its raw materials for industry, including all its oil. It imports some 80 per cent of its energy resources but has begun to develop a considerable nuclear power industry which already provides half of the country's electricity requirements. The country's considerable steel industry was, significantly enough, built up initially on its ship-breaking industry – the simplest and cheapest way of getting the raw material.

The country's population of over 21 million, boosted by the huge inflow of immigrants from the mainland at the end of the 1940s and in the early 1950s, speak several different Chinese dialects, but Mandarin is now widely spoken in this highly educated country; moreover, an important unifying factor among the population is their opposition to the Communist government across the straits in their homeland. However, labour unrest has developed in the country, especially since the government removed the ban on unions and strikes.

Like the government of South Korea, Taiwan's government has been strongly authoritarian in many ways; and although it does not interfere in most private industries, it does intervene through its state-owned enterprises in such key industries as shipbuilding and ship-breaking. Only since 1988, with the rise of a more democratic government, has any real political or economic liberalisation been taking place. Apart from allowing greater devolution of political decision-making, the government now allows Taiwanese to visit their homelands in China and even some reciprocal visits by Chinese from the People's Republic.

The economy of Taiwan is now typically dominated by the industrial and service sectors, but agriculture is still significant, though its percentage of the total labour force has been reduced from 47 per cent in 1965 to under 15 per cent today. With its good climate – a rainfall of 2,500 millimetres a year and temperatures in January and February rarely falling below 12°C – and 25 per cent of the land area available for agriculture, Taiwan has been at the forefront of agricultural research and development, especially into its main crops of rice, vegetables, sugarcane, sweet potatoes and bananas. Indeed, it is clear that Taiwan's success has been built on the classic sequence of agriculture first, then ISI for a short time, then EOI based on comparative advantages in international trade. Forestry is also important, but though forests still cover half the land area, the industry is inhibited by inaccessibility, by the poor quality of many of the trees, and by the government's conservation measures. Taiwan is indeed very conscious of the need to consider the environment in its economic development; and this is particularly in evidence in the Linyuan Industrial Zone in southern Taiwan where there has been a good deal of concern expressed about the disposal of industrial waste and its effects on marine life and the fishing industry.

But it is on industry that the Taiwanese economy depends most heavily, providing as it does over 85 per cent of export revenues, and engaging over 40 per cent of the work-force. One distinctive feature of Taiwan's industrial structure, however, is that it comprises largely small, family firms: there is nothing equivalent to the large *chaebol* of South Korea. While it still has heavy industries, on which much of its early success depended, there is now a shift towards high value-added activities, especially electronics and light industries. The older textile industry, too, is declining in relative importance. The move now is towards capital-intensive, high-technology industries, especially those with low energy requirements. Most of this new industrial growth is being located in government-selected sites, especially in the three export-processing zones of Taichung, Kaohsiung and Nantze.

Taiwan faces several economic challenges, one of which expresses what some observers describe as the excessively materialistic and acquisitive nature of the Taiwanese. 'The survival of the fittest' is said to describe accurately the ruthless business attitude of Taiwanese. Credit has been the fuel of the economy's growth. The Taiwanese view the stock-market as a casino and, after the stock-market crash of 1990, strict regulations have been introduced. Taiwan suffers from an excess of liquidity, and its foreign currency reserves are second in size only to those of Japan. Its accent on material success, often at the expense of more traditional values, affects citizens' attitudes and is of concern to the government. Moreover, it is associated with high wage rates, so that Taiwanese manufacturers, as elsewhere in the NICs, are increasingly moving offshore to lower-cost areas in the Philippines, Malaysia and Indonesia, and even across the straits to China. There are now some 3,000 Taiwanese companies operating on mainland China and one of Taiwan's leading textile manufacturers now operates one of Shanghai's largest government factories.

Taiwan is similar to the other NICs in being very heavily dependent on foreign trade. It has also had consistently heavy trade surpluses and has thus come up against the problem of international, especially United States' objections to its import tariff and non-tariff restrictions. The United States has recently persuaded Taiwan to appreciate its New-Taiwanese dollar to the level of the US dollar and to reduce barriers to imports. Disputes over intellectual property protection have also emerged and seem likely to dominate relations between Taiwan and other industrialised countries.

To deal with this over-dependence on foreign trade, Taiwan is now aiming to increase government consumption rather than trade, in the hope that this will drive growth over the next few years. Like Singapore, Taiwan views construction as a way of coping with market fluctuations and over-dependence on foreign trade; and it recognises the great importance of having an adequate infrastructural base. It is therefore embarking on an ambitious programme of infrastructural development to be carried out by domestic companies and by government. These projects include a mass rapid transport system for Taipei, the capital of 2.7 million inhabitants; improved telecommunications, railways and harbour facilities; and new energy projects. High capital outflows mean that these infrastructural developments will have to be paid for with external bond issues.

Looking to the future, it seems likely that Taiwan will attempt to bring about closer political and economic ties with Beijing. These will include increased direct trade across the straits with Fujian and possible integration with the Hong Kong–Southeast China–Taiwan 'growth triangle' referred to earlier. China is the natural source of supply of labour, especially for Taiwan's older labour-intensive industries, and is the natural market for Taiwan's products, given the common culture and language. Improvement in Taiwan–Chinese relations is important to Taiwan, not only for economic reasons, but also because Taipei wishes to reduce international tension by removing any military threat from the mainland, and because Taiwan wishes to be accepted into the international community via the United Nations. Taiwan has now set up a development fund for developing countries in the region.

Conclusions

This is not the place to examine whether there are any lessons to be learned from the experience of Japan and the four NICs, or whether this experience suggests some kind of model for other less prosperous countries to follow in their own developmental paths. The differences among the five economies are great and sometimes profound, in area, physical environments, population size and levels of homogeneity, types of government, historical legacies and in a whole range of internal and external political and economic problems. What they all share seems much less remarkable: very few, if any, natural resources; an early emphasis on export-oriented rather than import-substitution industrialisation; and an increasing tendency to engage in offshore manufacturing operations in the lower-cost locations available in the less prosperous countries of the region, including China. In all cases, too, governments have been selectively interventionist, sometimes directing but more usually guiding the market, and always focusing on the need for growth.

Nevertheless, it is clear that many, if not most, of the conditions which allowed these countries to pursue successful economic strategies were unique to the time at which their development took place. External conditions, including especially expanding world trade; pressure for specific policy reforms from the United States; weak domestic labour movements; and an absence of leftist or populist parties – all these circumstances were decisive. The very different circumstances of today suggest that any idea of a West Pacific Rim model, based on the experience of Japan or the four NICs, is unlikely to bear close examination.

9

China since 1978

As one of the poorest, and at the same time the largest and most populous country in the West Pacific Rim, China might seem to require rather different treatment in any attempt to explain its 'success'. Official written and statistical material on the progress of China's economy since the Reforms of 1978 are widely regarded as uniquely suspect, and scholars point out that much of the Chinese-language literature on China's economy is little more than government propaganda and that official statements and statistics must be treated with more than a little caution. The economy of this communist, centrally planned state is still weak, inefficient and corrupt.

All this is true. But it is still worth attempting to explain economic growth in China because, though it is still relatively poor, the country's economy is showing significant improvement and in some areas, notably in southeastern China, there are even signs of the kind of 'miracle' development normally associated with the four NICs. According to most observers, living standards have tripled over the last decade or so (Yahuda, 1990, p. 55). But, more importantly, as pointed out in Chapter 1, China's rate of growth, at over 10 per cent a year since 1978, is the highest in the world. It is further predicted that China may well have the fastest growing economy in the whole of Asia during the 1990s. Perhaps it is now not unreasonable to consider China as a potential NIC, and even, within a few decades, Asia's 'next economic giant' (Perkins, 1989), a prospect which has profound global as well as regional implications.

It was suggested in Chapter 1 that China is particularly well endowed with natural resources. A vast country, with a wide range of contrasting environments, China also has a remarkable array of mineral resources for industry. Apart from significant deposits of iron ore, bauxite and many other minerals, China has rich energy resources: oil and natural gas reserves are immense and the country is already the world's sixth largest producer. But easily the major energy resource providing power enough to last 400 years, and China is currently the largest coal producer in the world. Hydroelectric power, especially in the wetter south, is another major source of energy, and increasing supplies of power are now beginning to be developed from geothermal fields and from bacteria-produced biogas. Finally, the potential for energy from nuclear power is believed to be immense.

Yet while China's industrial resource base is very striking, this has to be set against the huge population. As Edmonds (1991) points out, the per capita resource base in China is in fact below the international average. Moreover, the resource potential has yet to be fully exploited. Coal resources are still inefficiently produced and distributed; and much more foreign investment and technology is required to develop the hydroelectric potential and to search for and exploit new oil fields, especially offshore. The role of China's natural resources in industrial development is indeed more limited than might be expected. Resources for industry are heavily constrained by poor transport, lack of capital and technology, limited investment and restricted market opportunities. Nevertheless, the potential is there and, with the kind of joint ventures such as that now taking place in the new coal mines at Pingshuo, there is every reason to expect China's natural resource base for industrial development to improve still further.

China's human resources are vast. Her population of 1.2 billion represents about one-fifth of the world's total population and provides a vast potential market for domestic and foreign producers as well as easily the largest labour force of any country. Moreover, with the rate of growth of population now as low as 1.4 per cent per annum – reflecting the success of such measures as the 'one-child-per-family' policy – there is little danger of China's economic growth rates being overtaken by population increases (Croll, Davin and Kane, 1985). Furthermore, as Table 1.4 shows, the population is relatively healthy, with very low infant mortality rates and reasonably high life expectancy rates (Bannister, 1987). The main problem about the quality of the population in the context of economic growth is what some commentators regard as the 'passivity' and 'demoralised' nature of the Chinese population, especially in urban areas; but this kind of subjective comment is usually coupled with remarks about the repressive nature of the Communist government under which the Chinese have to live.

Changes in government policy

Whatever success China's economy has achieved since 1978 must be put down to changes in government policies. The present government of China, subjected to a great deal of opprobrium as a result of the events in Beijing's Tiananmen Square in June 1989, is certainly Communist, authoritarian, repressive and non-democratic. It remains in almost complete control of the economy, operating within a tightly controlled political and administrative system, and it seems determined to maintain China's position as the last great communist state. Yet in late 1978 the government made a significant break with the past and introduced what it terms 'market socialism', the purpose of which is to modernise China's economy and improve the standard of living of its peoples. It has attempted to reinstate markets as a central feature of socialist economic theory and practice. This means the continuing role of the 'law of value':

> the operation of supply and demand, the Marxist 'economic categories' (such as price, profit, interest, commodity) which Maoist theorists had seen as 'bourgeois' notions, and 'the law of distribution according to labour' which demands unequal pay for unequal work, challenging Maoist egalitarianism and underpinning new wage systems using piece-rates and differential bonuses. (White, 1988b, p. 87).

The ownership system has been diversified, with private enterprise gaining some ground, especially in the commercial and service sectors. But state ownership remains dominant, even though decisions and control of operations may lie with the individual.

More generally, the government has been moving from ISI to EOI, though cautiously and slowly. Furthermore, there is renewed emphasis on agriculture in an attempt to deal with the need for improved food supplies, especially to the urban–industrial areas, and to increase the purchasing power of the rural population. The government has also encouraged a shift away from heavy industry in favour of light industry. In all this there is constant emphasis on labour incentives. But perhaps above all, the Reforms since 1978 have included the government's 'open-door' policy of encouraging foreign investment and technology, together with the setting up of fourteen open-door coastal towns and five Special Economic Zones into which foreign investment and semi-capitalist or capitalist enterprises have been channeled.

This limited liberalisation of the economy has resulted in a considerable boost to China's economic fortunes, but by the mid-1980s the economy was showing signs of overheating with growing inflationary pressures in evidence. In 1988, therefore, the government introduced an austerity or retrenchment plan to cool the overheated economy and to reduce inflation. More specifically, the government hoped to reduce the overall rate of growth to well below 10 per cent, to cut inflation, to control investment, to bring demand and supply into something like equilibrium, to increase farm output, and to increase energy supplies. There was also some further decentralisation of decision-making powers, especially in the southeastern provinces of Guangdong and Fukien, where four SEZs are located.

Agriculture

In this vast and heavily populated country it is inevitable that agriculture should for some time continue to be of central concern in the government's economic policy. Four-fifths of the population are still rural and 77 per cent of the labour force is still engaged in agriculture. Thus while the government is clearly determined to make industry the leading edge of its economic policies, it must continue to support development in the agricultural sector in order to feed its population, both rural and urban, and to provide the necessary raw materials for some of its industries.

However, China suffers from two fundamental ecological constraints which continue to impose limitations on the country's agricultural growth. The first of these is the serious shortage of good, arable or workable land. Although China is an immense and ecologically varied country, only 10 per cent of its land surface is cultivated; moreover, it is widely accepted that the scope for increasing the cultivated area is negligible, and that increases in production will have to come from the intensification of production on existing agricultural land. This is a particularly critical problem, because the country's huge population of well over a billion means that the density of population per unit of cultivated land is already very high at 0.10 hectares of cultivated land per capita. Moreover, over 90 per cent of China's population lives in the southeastern third of the country.

Climate, and especially a lack of water, is the second major constraint facing agriculture in China. Much of China north of the Yangtze River is a moisture-deficit area with less than 1,000 millimetres of rainfall a year. Moreover, seasonality and variability of rainfall become more marked northwards and westwards, making even parts of the best land 'marginal' from time to time. There is also the fact that China suffers from serious climatic hazards – droughts, floods and typhoons – which have always plagued China's territory, more especially in the eastern river valleys, flood plains, river basins and deltas where so much of the country's population lives.

After the Reforms of 1978 the return to family farming under the Responsibility System, the abandonment of the commune system, and the use of the market mechanism initially improved productivity in agriculture. Within six years rural incomes doubled, creating in the countryside a solid base for further reform and growth. But by the mid-1980s it was apparent that productivity was stagnating and that further substantial investment in agriculture was needed if sustained improvements in productivity were to be maintained. Since then agricultural productivity has increased again, largely as a result of increased investment of some 15 per cent in 1988 and, in 1989, by raising the purchase price of grain by 18 per cent. Other administrative measures have also been adopted, but China's agriculture still has some way to go if it is to provide a sound basis for the programme of urban–industrial development on which the government is now embarked.

Industry

Before 1978 industrial policy in China was broadly one of import-substitution, emphasising heavy industry and utilising labour-intensive methods wherever possible. Since 1978, however, what has been called the 'reform paradigm' has replaced the essentially Soviet-style planning system. There is now 'a re-orientation towards light industry, agricultural and consumer goods, international ties, the diversification of ownership and the reinstatement of markets' (Benewick and Wingrove, 1988, pp. 5–6). This, so it is argued in China, involves moving towards market socialism.

This change in industrial policy has certainly achieved many of these aims. The value of output from heavy industry is now only marginally more than for light industry, and this trend is continuing; international links, including especially foreign investment and joint ventures, have grown markedly; and market forces are already having their effect on industrial productivity. Most fundamentally, perhaps, the government has been successful in cooling down an overheated economy by reducing the economic growth rate to about 7 per cent.

However, the industrial structure is still dominated by over one hundred thousand state-owned enterprises, most of them in urban areas, which account for 77 per cent of the industrial output. Although there are some three hundred and four thousand collectives or co-operative enterprises, these are mainly in rural areas and account for only 22 per cent of the total industrial output (Yahuda, 1990, p. 56). Furthermore, the austerity or retrenchment measures of 1988 hit these smaller enterprises, whether collective or private, located mainly in the new townships in the rural areas. Retrenchment among these enterprises also had a damaging effect on the rural peasants who regard work in these 'rural' industries as providing an enhanced source of income.

While there has been some shift away from the former rigid, centralised system of administration, and private industrial enterprises are now allowed, state control and state ownership is still very evident. And although the Reforms have to some extent begun to undermine the old administrative controls of a centralised system, the necessary macroeconomic controls have still to be put in place. Administrative and economic controls are not yet separated. As one authority puts it, 'China is a half-way house between a communal Stalinist economy and a reformed market one: the country suffers the worst of both worlds' (Yahuda, 1990, p. 55).

One of the major problems the government has had to cope with is price reform, essential if the market mechanism is to work effectively, but difficult to introduce painlessly into an economy not used to it. A particular problem arose as a result of the introduction in 1985 of a two-tier price structure. It was designed to stimulate output based on high productivity, allowing economic units to sell 'surplus' output – that is, output beyond that demanded by the state – at prices higher than those contracted by the state. However, many organisations simply supplied each other's needs at the lowest price, thereby ensuring supplies and evading tax simultaneously. The scheme also encouraged corruption, for organisations then sold at higher prices supplies acquired at the lower price.

A further problem the government has had to face in its attempts to move some way towards a market economy is how to deal with the system of subsidies, especially for the state-run industries in which most of the 100 million urban work-force are engaged, though it also applies to loss-making state farms in the rural sector. In a market economy it is clearly unacceptable for workers to be paid salaries and even bonuses, even when they are not working; and it is accepted that subsidies can so easily inhibit the kind of attitude a market economy – even a socialist market economy – demands. It is now hoped eventually to eliminate these subsidies altogether and to carry out wage reforms to link wages with economic performance.

The open-door policy

But far more important in trying to explain China's recent economic growth is its open-door policy (Bucknall, 1989). This affects the most crucial element in the government's policy for rapid and successful industrial development: foreign investment, including joint ventures (Pomfret, 1990). Since 1978, the government has encouraged foreign investment, channelling it mostly into the eastern and southeastern coastal regions. This concentration of foreign investment and capitalist or semi-capitalist industrial activity was motivated by several factors, including ease of access and the multiplier effect; but it was also moti-vated by the desire to protect central and western China from the less desirable aspects of contact with the West. In fact, though there is large-scale investment by the West (in east and southeastern China) and by

Japan (especially in the Liaoning region of the northeast), Hong Kong and Taiwan are the most active in foreign investment. Already, individuals and companies from Taiwan are operating joint ventures and investment in Fujian, and direct trade of over $2.8 million is now carried on between the two countries, and especially between Taiwan and Xiamen SEZ. Hong Kong's investment is mainly in Guangdong, about which more will be said later. And all this foreign investment is supported by loans from the World Bank, the Asian Development Bank, and from the Japanese Overseas Development Agency (ODA).

The open-door policy, allowing the kind of foreign investment just referred to, was the direct international complement of the Reforms of 1978. It provides for a cautious welcome for foreign firms operating in China and for a degree of free trade. It was designed to encourage foreign investment and the kind of technology transfer necessary to develop China's exporting capability. But it was also expected that the policy would encourage outward-looking habits among the Chinese and subject them to the price disciplines associated with world market operations.

The policy was applied selectively to specific geographical areas in eastern and southeastern China: to fourteen coastal open cities (like Shanghai and Shandong), with extra incentives and tax breaks for foreign trade and investment; and to the five SEZs stretching from Hainan in the south, through the three SEZs in Guangdong Province, to Xiamen in Fujian Province opposite the island of Taiwan. These SEZs provide even more attractive sites for foreign investment, trade and technology.

The southern 'miracle'

Resulting directly from China's open-door policy, the two southern provinces of Guangdong and Fujian are now experiencing the kind of rapid economic growth that is attracting the description 'miracle'. At least in this part of China there is a boom economy, with per capita GDP incomes of around $2,000. Guangdong is a large province, containing over 63 million people. Its industrial output of clothes, shoes and toys rose 15 per cent a year during the 1980s. Exports rose similarly, and Guangdong accounted for one-third of China's total exports by value in 1990 (Vogel, 1990). As for Fujian, its success has been even more dramatic, because it began from a much lower base. Fujian in the 1970s was one of China's poorest provinces, yet today it is second only to Guangdong as the destination for foreign investment. Indeed, Xiamen is the fastest growing SEZ in China, achieving annual increases in industrial growth of up to 40 per cent (*The Economist*, 1991c, pp. 24–6).

However, it is important to recognise that wealthy Hong Kong and Taiwan have become integrated economically with their respective continental hinterlands Guangdong and Fujian. The cost of labour is clearly an important consideration: labour costs five times as much in Hong Kong as in Shenzhen, the SEZ just across the border, and ten times as much as elsewhere in Guangdong. Land, too, is much cheaper in China: in Guangdong land costs are only 2 to 3 per cent of costs in Hong Kong; in Fujian land costs are about one-tenth of what they are in Taiwan. Nor does this mean any drop in quality. The productivity of workers in southeastern China is, according to most authorities, comparable with that found in Hong Kong and Taiwan. All these factors, together with the small amount of bureaucratic interference, especially in Guangdong, have contributed to the kind of rapid and successful development the rest of China will wish to emulate.

It is tempting to speculate about the possible future political as well as economic implications of these developments in southeastern China. One authority writes:

> Imagine that the Chinese empire run from Beijing were to disintegrate – Guangdong, Fujian, Taiwan and Hong Kong could form the Republic of Southern China. This new country would have a population of 120 million and a combined GDP of roughly US$310m, which would put it on a par with Brazil. (*The Economist*, 1991b, p. 18)

This may be a quite unreal scenario, but certainly the pattern of industrial development and the differential growth rates of productivity and prosperity seem likely to create increasing stresses on the unity of the state.

The Beijing government is of course well aware of this problem and is in various ways attempting to redress the balance between the booming south and the rest of the country. For instance, an attempt is being made to reinstate Shanghai as the country's major international financial and business centre. Shanghai is receiving much-needed government support for its development projects, notably the Pudong Development Project just to the east of the city. This involves developing 350 square kilometres of farmland and swamp into an industrial, commercial and financial zone. Pudong aims to become a marketing, service and financial centre and a catalyst for the economic advancement of the whole Yangtze River region, as well as an effective counterbalance to economic success in the south (*FEER*, 1990a, 1990b).

Conclusions

China is a good example of a country where any analysis of its economic development, either current or future, must be underpinned by political matters. This is not just because it is a communist country, with its predictable theory and practice, involving politics in every aspect of economic life. It is also because China is now the last great communist state, and it cannot avoid looking over its shoulder at what has happened to the former Soviet Union and the countries of Eastern Europe. The point from which the Chinese government starts is that China will remain a communist country and that its state identity will remain firm. There is to be no *glasnost* in China. While the government is ready to consider limited and controlled liberalisation of the economy, such liberalisation must always operate within the limits set by the political imperatives. As noted in this chapter, the government has recognised the need to improve food production, to cool the overheated economy, and to control inflation. And it is prepared to follow the advice of Western economists in dealing with these problems – but only if their political control remains inviolate.

The overriding need, as far as China's immediate economic future is concerned, is political stability. A sudden democratic revolution must always remain a possibility. Yet foreign investment, technology transfer, and all the paraphernalia associated with a large command economy moving into closer relations with the world economy will always depend very much on confidence in China as a stable political unit.

Some writers believe, or at least seem to imply, that China's main problem lies with the present political system, that unless China embraces democracy it cannot continue to prosper economically. But this ignores the fact that all the governments of the more succesful NICs have had or still have strong, *dirigiste* governments. Rapid and successful industrialisation in the NICs is not and cannot be correlated with non-interventionist governments. It has already been pointed out that state direction and control has been a critical factor in the industrial growth of Taiwan, Singapore and South Korea. The problem is rather whether the Chinese government is prepared to adapt the nature, degree and style of its intervention so that it guides rather than rigidly controls the economy. There is plenty of evidence that economic imperatives as well as current theoretical and ideological arguments in China are being used to justify this kind of change or adaptation.

The Chinese no longer argue that communism and capitalism are logically incompatible. In 1990 Li Peng argued that planned and market economies each have their own strengths and weaknesses and that China needs to draw upon the strengths of both in order to build a mixed, integrated economy. However, even if the government is not prepared or able to change the nature of its political control over the economy in this way, there seems little doubt that the demonstration effect of successful foreign firms and businesses operating in China will in the end force the government to go along with the principle of the

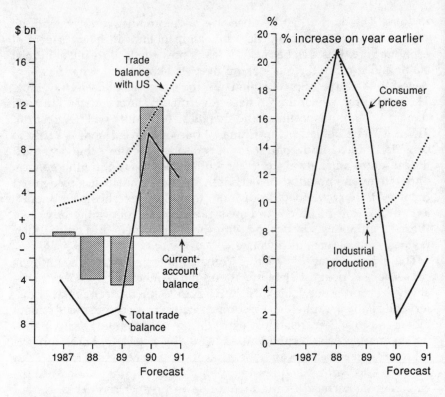

Figure 9.1 China: trade balances and changes in consumer prices and industrial production, 1987–91 (Sources: Smith New Court; EIU; US Commerce
Department; DRI/McGraw Hill; *The Economist*, 30 November, 1991)

'guided' or 'governed' market, the mixed economy referred to by Li
Peng. Economic growth, in other words, may well come as much from
pressure from below as from changing policies at the the top.

In spite of China's present low position in the economic growth
league (Table 1.1), there are firm grounds for taking an optimistic view
of China's economic future (Figure 9.1). Foreign trade, investment and
technology are already providing the catalyst, as they did in those other
East Asian economies China is now trying to emulate. China's physical
and human resources are potentially vastly greater than in any of her
neighbours, and there can be no doubting China's determination to
achieve economic success and global economic power. While political
imperatives may reduce the speed at which development occurs, there
is no reason to believe that the dead hand of the state will continue to be
anything like as damaging as some writers suggest. Even if one assumes
the continuance of the present government, it is clear that this need
not be an insurmountable barrier to growth. State industry is being

reformed from within, but in a way and at a pace determined by political imperatives. Domestic private industrial enterprises are boosting the country's economy, helped incalculably by the inflow of investment, firms and technology from abroad.

There are many imponderables. One of these is the issue of Hong Kong. Another is how far economic freedoms are consistent with the present tightly controlled and repressive political system. Much of the literature in the West suggests that economic liberalisation and a repressive, totalitarian political system are incompatible, though in China there are many younger academics who see no logical reason why this should be so. Looking to China's economic future, indeed, it may well be that the worst-case scenario is not that Communist control will continue, but that the clamour for political freedoms will ignore and then destroy – as they have in the former Soviet Union – the economic progress on which the prosperity as well as the stability of the Chinese state must ultimately depend.

10

The Southeast Asian sector

Any attempt to explain the economic success of the countries of the Southeast Asian sector of the West Pacific Rim must take into account the fact that their differential levels of success are very marked. Even leaving aside Singapore, discussed earlier as one of the NICs, the countries dealt with in this chapter range from Brunei, a high-income economy with a per capita income rivalling that of Japan, to Vietnam and Indonesia, both of which are still classified as low-income economies. In between are Malaysia and the Philippines, classified as lower-middle-income economies (Table 1.1). 'Success' in this part of the West Pacific Rim, then, is a very relative term and, as with China, it is only within the recent historical context of each country's economic growth that explanations of success can have much point.

Nevertheless, as indicated in Chapter 1, each of the constituent states of this region has begun recently to share in the increasing prosperity of the West Pacific Rim as a whole, and most predictions suggest that by the early part of next century several of the economies of the Southeast Asian sector will have reached the status of NICs.

The discussions presented in Chapters 1 and 2 are particularly relevant to the economies examined in the present chapter. They are all former colonial territories; they are resource-rich; and they operate a mainly colonial pattern of trade, especially so for their exports, which are still dominated by primary agricultural and mineral commodities.

Vietnam

Lying immediately to the south of China, Vietnam comprises two well-populated deltas, the Red River delta in the north and the Mekong River delta in the south, joined by a sparsely populated, narrow, mountainous coastal strip. It is hardly surprising, then, that the country has experienced difficult problems in its efforts to achieve national unity, even during its colonial period, when the French ran their territory as three units. After the war the country experienced a number of convulsions and wars with the French and then with the United States; it was not until the end of the Vietnam War in 1975 that Vietnam was united under the control of North Vietnam. The following years, however, saw further conflict: with Cambodia, which Vietnam invaded in 1978, and with China, which invaded Vietnam in 1979.

Since 1975, and especially since Vietnam's invasion of Cambodia in 1978, Vietnam has been largely isolated internationally. Part of the reason for this isolation – and the reason, incidentally, that there are so few reliable statistics about its economy – is that while North Vietnam aligned itself with the former Soviet Union, Southern Vietnam aligned itself with Communist China and the West. Furthermore, Vietnam is still perceived to pose a threat to regional security because it retains the largest and most powerful army in Southeast Asia: 1.3 million men under arms, costing the country some 30 per cent of its total national income. Southeast Asian (ASEAN) countries, as well as the United States and the West, have isolated Vietnam, economically, politically and strategically.

At unification in 1975 North Vietnam imposed on the reunified country its own rigid brand of Stalinist socialism, operating the economy under a highly centralised communist five-year plan system. The immediate result was to increase poverty and accelerate a decline in the country's economy. All 'capitalist remnants' in the south were eradicated; some four hundred thousand Saigon inhabitants were forced out to work in unworkable 'New Economic Zones'; and coercion was employed to implement a programme of collectivisation in the south. But by the mid-1980s it was apparent that something had to be done to halt this decline. In 1986, therefore, Vietnam has adopted more pragmatic, reformist economic policies (*Doi Moi*). The collectivisation of agriculture was stopped, more reliance was placed on the market, and private ownership of land legalised. Private enterprise in industry was encouraged and foreign investment welcome. Despite some set-backs, such as the serious drought of 1987/88 and serious inflation, reckoned to be around 700 per cent in some years, there are now clear signs of improvement. As noted in Chapter 7, Vietnam's market-oriented policies in agriculture have already resulted in the country becoming an important rice exporter. More recently, the government has put forward a draft strategy of socioeconomic stabilisation and development up to the year 2000 which is designed to double the per capita income, increase rice production by 50 per cent, triple the electricity output, and increase exports fivefold by the year 2000. However, if Vietnam is to achieve these aims it has at the same time to encourage a number of other changes, including more effective birth control: the present population of 65 million is otherwise likely to reach 85 million by the year 2000 (Table 1.3).

Vietnam is of particular interest to the discussion partly because it illustrates two issues that arose in earlier chapters. First, Vietnam is by some criteria the poorest country in Southeast Asia (Table 1.1) and, indeed, is among the ten poorest countries in the world. Yet it has an excellent natural and human resource base. The mineral resources for industry are mainly in the north and, according to most writers, the northerners provide the most disciplined work-force with a long tradition of entrepreneurial ability. 'Given the superb natural and human resources available, the country's poverty is all the more shocking and embarrassing to Vietnamese leaders' (Neher, 1991, p. 168). Explanations of this relative economic failure must, as in China, include the country's experience of communist, centrally planned economic policies. But they must also include in Vietnam's case the devastion wrought by a series of wars, the sudden cessation of United States involvement, and serious mismanagement by its leaders (Neher, 1991, p. 169). Secondly, Vietnam mirrors the problem now being faced in China: how to move from a communist, centralised economy to a more market-oriented economy while at the same time holding on firmly to political control. As with China, the Vietnam Government refuses to consider political reform while encouraging economic reform. At a hardline political meeting in 1989 the Government's determination to reject any form of 'political pluralism' was firmly restated. However, it may be that change within Vietnam will come from the old north–south tensions. The disaffected military in southern Vietnam have put up with rule from the north for sixteen years; and throughout the country there are now many who believe that only when communism has been removed will the country's economy make the necessary advances.

Nevertheless, economic imperatives require Vietnam to break out of the international embargo placed on it by ASEAN, the United States and China. There have indeed been some improvements since 1989, though the country still suffers from its pre-industrial structure, its lack of capital and investment, poor education and a weighty, inefficient bureaucracy. Now that Vietnam can no longer rely on help from the former USSR, it may well be that the key to its future lies in two directions: how far it can accept the social and political consequences of economic reforms; and how far it is prepared to agree to the domestic and foreign economic changes being demanded by international agencies like the International Monetary Fund.

The Philippines

A vast archipelago of over seven thousand islands, the Philippines is in several ways unique. Until it came under the control of the United States in 1898, the country had been under the control of Spain for over four centuries and is now the only largely Christian country in the Far East, with 85 per cent of its 63 million population professing the Catholic faith. However, this has not proved to be any kind of unifying force in this very fragmented state. Apart from the many different language groups, there are significant numbers of Muslims, particularly in Mindanao, and a small but economically powerful group of Buddhist Chinese. Politically, too, there are many factions in this multi-party democracy and a continuing security threat from the Communist insurgents.

Indeed, in discussing the Philippines it is tempting to focus not so much on explaining its success as on finding reasons for its relative failure. Even though China, Vietnam and Indonesia are lower down the per capita income league, the Philippines' economy has shown negative growth in several years in the 1980s, and it is only in the last few years since 1989 that any evidence of 'success', however limited, can be found (Table 1.1).

It is not that the Philippines is without advantages. It is a rich country in terms of natural and human resources. Its minerals include copper – in which the Philippines is Asia's largest producer – gold, silver and coal. It also has the climate and, in most cases, the soil to support a flourishing agriculture. But perhaps above all, the population, though poor, is by any standards well-educated, largely as a result of the school and college system set up during the period of American control: the Philippines now has 26 per cent of its college-age population in higher education – more than in the United Kingdom. The population is also politically aware and always ready to engage in political debate.

In trying to achieve more rapid economic growth, however, the government has had serious difficulties to face. Apart from the centrifugal tendencies mentioned above, the society is, according to some writers, particularly prone to violence and to indebtedness: 'perhaps nowhere in Southeast Asia is the populace so articulate, and so fond of participating in political debate'. Yet 'the spirit of indebtedness remains as hard to break as does the spirit of violence' (World of Inf. 1990, p. 199). Indebtedness is also a characteristic of the country's economy: in 1991 it was estimated to be almost £30 billion, costing the country over 40 per cent of its national budget to service. Endemic corruption is another problem, usually referred to as 'crony capitalism', reflecting the way in which the previous president, Ferdinand Marcos, placed his own cronies into positions of power. In the circumstances of political instability, too, there has been a flight of capital from the country by the richer Filipinos, only partly compensated for by remittances from the many Filipinos, including over one hundred and seventy five thousand women, working abroad. The country is also about to lose heavily in financial terms from the removal of United States military bases in the country.

There is still a large element of a colonial economy about the Philippines. It is still very much an agricultural country, dependent on commodity earnings from primary exports for 30 per cent of its GDP, and for 40 per cent of its export earnings. It is hoped that a new land reform programme, introduced in 1988, will provide a further 2.8 million hectares of agricultural land for some 2 million landless farmers over the next ten years. But the Philippines' agricultural economy must for some time continue to suffer from fluctuations in demand and prices for its major agricultural exports as well as for its lumber and copper. Unless land reform succeeds in revitalising the country's agricultural sector, the Philippines will not be able to provide the sound agricultural base on which its industrial growth can be built.

Industry accounts for 25 per cent of GDP but still involves mainly the processing of primary products in largely government-owned companies. The emphasis been shifting from ISI to EOI since 1972, when the first Export Processing Zone (EPZ) was established, yet the country still operates an unacceptably high rate of import protection. Since 1986 foreign investment has increased by 20 per cent, construction by 16 per cent and manufacturing by 7.4 per cent. One of the country's major needs is for a greatly improved infrastructure: it suffers from serious bottlenecks, all of them compounded by the country's physical fragmentation. The country's industrial future also demands much greater foreign investment. As indicated in the discussions on several of the other West Pacific Rim countries, the Philippines are increasingly being used for offshore operations, especially from the NICs, whose investment in the Philippines already exceeds that of Japan.

As in most of the countries in the Southeast Asian sector, it is politics that hold the key to further development in the Philippines. Whether the present democratic government under Ramos will continue, in the face of numerous coup attempts, the discontent of the army, the continuing threat from Communist insurgents, and a still faltering economy remains to be seen. Yet without political stability, the investment so urgently required is unlikely to be forthcoming.

Malaysia

The three units that today make up the Federation of Malaysia – Peninsular West Malaysia and the states of Sarawak and Sabah on the island of Borneo – were all formerly under British colonial rule. In 1963, together with Singapore, they combined to create the Federation of Malaysia, from which Singapore seceded in 1965. Hot and wet throughout the year, the country is still heavily forested, especially in the mountainous interiors. Even more than Vietnam and the Philippines, Malaysia is a remarkably resource-rich country. Its natural resources include oil and natural gas (11 per cent of export earnings), tin (the world's largest producer), iron ore, bauxite and copper, as well as tropical hardwoods. These natural resources are unusually well balanced against a fairly small population (17 million) and large land area (330,000 square kilometres – about the same size as Vietnam). Indeed the only real problem Malaysia has come up against in its recent development of her natural resources is international concern over the environmental implications of her forestry industry, especially in Sarawak.

Malaysia's human resources, however, provide the country with some difficult problems. The population of 17 million, 82 per cent of whom live in Peninsular Malaysia, is still increasing at a high rate (2.8 per cent); migration into the major urban centres like Kuala Lumpur seems to be out of control; and there are many small indigenous groups like the Iban (in Sarawak) and Kadazans (in Sabah) which are only nominally Malay. But by far the major problem in Malaysia is the existence of the Chinese (35 per cent) and Indians (9 per cent), groups which to some extent are distinguished geographically and occupationally.

Malaysia's New Economic Policy, established in 1971, is in effect a policy of positive discrimination in favour of the Malays. It is usually referred to as an experiment in social engineering, being an attempt to redress the social and economic inequalities of a culturally divided nation, nearly half of them living below the poverty line. In 1971 Britain 'still owned 60% of the country's corporate wealth; the ethnic Chinese dominated the booming local commercial sector; and the Malays (Bumiputras) were still predominantly on the land – focus of much of the poverty – or in the Civil Service' (*THES*, 1991, p. 10). Central to the New Economic Policy was the establishment of quotas for business and higher education to create a pool of Bumiputras to step into managerial and professional roles. Today, over twenty years later, less than 20 per cent of the population are living below the poverty line; there is now a substantial Bumiputra middle class; and about 75 per cent of the country's wealth is now Malaysian owned, about a fifth of which is in the hands of Bumiputras. The New Economic Policy, with its relatively modest proposals and its awareness of the need to retain the economic support and expertise of the Chinese, has played a significant role in defusing social tensions and encouraging joint efforts in development planning.

The structure of the economy today is changing rapidly, but still emphasises the production of raw commodities like oil, natural gas and tin, or such basic agricultural commodities as rubber, palm oil and cocoa. Agriculture provides 21.5 per cent of GDP and involves 32 per cent of the labour force, and its importance is clear from the fact that Malaysia produces 35 per cent of the world's rubber and 60 per cent of the world's palm oil. Nevertheless, it was soon realised by the Malaysian government that it could not hope for rapid sustained growth on the basis of commodity exports alone.

Industrialisation is now developing successfully and its contribution to the GDP is already just ahead of agriculture. Early development plans emphasised ISI for consumer goods, but after 1981 the emphasis changed to EOI and the growth of heavy industries, like steel, cars and cement. These industries, however, have faced a number of problems on the world market and light industries are now expanding more rapidly. Malaysia is already the world's third largest manufacturer of integrated circuits and the largest exporter of semiconductors. The country is now well on the way to becoming an NIC, though its Prime Minister is reluctant to accept that status because to do so would strip the country of its 'developing country' duty-free trading privileges, subject its exports to quotas, and force Malaysia to increase the value of its currency. This attitude, a combination of nationalism, xenophobia and something approaching paranoia, is widespread and reflects many issues, including the desire to curb the power of the Chinese, and indeed of all 'foreigners'. The government is, however, trying to stimulate the private sector by selling off over half of the state-owned companies it currently holds. Nevertheless, it is beyond question that the government continues to play a very significant, if not decisive, role in economic policy.

Theoretically, Malaysia's economic future should be bright, with its excellent natural resources, a booming economy (7–8 per cent a year growth rate), and an excellent infrastructure. There is already substantial foreign investment, and relatively low costs have encouraged companies in the NICs and Japan to move their operations to Malaysia. Apart from Singapore, Malaysia is easily the most successful economy in the Southeast Asian sector of the West Pacific Rim and most writers are optimistic about Malaysia's future prospects. Certainly the government is following a very pragmatic economic policy, emphasising the need to create a market-oriented economy and introducing the privatisation of public utilities, communications and transport. A 'Look East' policy stresses the adoption of the work ethic and methods of Japan and South Korea and increased trade with its Asian neighbours (Neher, 1991, p. 115). Malaysia's foreign exports structure is particularly promising because it shows great strength in depth: while its biggest export earner is oil and natural gas, this represents only about 11 per cent of the country's export earnings, because so many other items also appear in the pattern of exports. With the example of Singapore on its doorstep, Malaysia is grasping enthusiastically at its opportunities for rapid economic improvement. This is all the more remarkable in view of its domestic communal problems, internal factional fighting and its apparent lack of commitment to capitalism and the free market. Once the emphasis on trying to achieve social equity, as expressed in the New Economic Policy, ceases to preoccupy government, Malaysia's economic growth should begin to rival that of the NICs. The New Economic Policy has in fact now been replaced by the more confident National Development Policy, which is less obsessed with social equity and faces up to the new realities involved in sustained economic growth.

Brunei

The former British protectorate of Brunei, which gained its independence only very recently in 1984, is an exceptional, indeed unique, case not only in Southeast Asia, but in the entire West Pacific Rim. In terms of per capita income it is one of the richest, possibly *the* richest, country in the world and as such falls into the category of a high-income country according to World Bank criteria. It is a small but very wealthy country dependent on its oil and natural gas production, with no large numbers of rural poor, and no significant problem of rural–urban migration. However, it is difficult to regard Brunei as developed because it is in no sense an urban–industrialised country.

Brunei comprises two tongues of land on the coast of Borneo between the two East Malaysian countries of Sarawak and Sabah. Its land area is 70 per cent primary or secondary tropical rain forest. Its population of only three hundred thousand, over half of whom live in or around the capital of Bandar Seri Begawan, is largely Malay, though there is a significant Chinese minority of about 28 per cent. The standard of living is high, with free medical services, education and pensions, and there are government subsidies for food and housing. In spite of high rates of population growth (3.7 per cent annum in 1989) and a bottom-heavy age structure (46.1 per cent aged 20 or less), its standards of health and education are exceptionally high. The government has also invested heavily in infrastructure, including a new airport, a deep-water port, good roads, hospitals, schools, a university and numerous impressive government buildings.

There are two main points to make about Brunei for our purposes. First, it is the classic example of a country whose wealth is based on a natural resource, oil and natural gas, which accounts for over 97 per cent of its exports by value. Secondly, it is the only example in the West Pacific Rim of a country which has not tried conventional economic development in either agriculture or industry. While in theory Brunei has a democratic constitutional government, in practice the Sultan of Brunei is the absolute ruler. Nevertheless, the government has followed a liberal trade policy. The current development plan encourages a shift towards agricultural development and import-substitution industrialisation over a wide range of products. The reason for the shift to agriculture is partly that the country has to import over 80 per cent of its food requirements – agriculture, indeed, accounts for only 1 per cent of GDP – and the government aims to achieve greater food security by, for instance, developing mechanised rice projects and by a policy of diversification which anticipates the possible decline of the oil and gas reserves in some twenty years time. In the industrial sector, the government is adopting a conservation policy, controlling the level of exploitation and exports of its oil and natural gas reserves.

Indonesia

A vast, sprawling country of almost fourteen thousand islands and covering over 1,919,000 square kilometres, Indonesia was formerly a colonial possession of the Dutch (the Dutch East Indies). Like Malaysia, Indonesia is resource-rich, but it has a very much more fragmented and extensive land area and a population of 175 million, over ten times the population of Malaysia. Its major natural resources are also very similar, oil and natural gas easily dominating. But the population of Indonesia is very much more homogeneous than that of Malaysia: 96 per cent are Malay of some kind, but there are very many small non-indigenous groups including the Chinese who, comprising only 3 per cent of the total population, control an estimated 75 per cent of corporate assets. Culturally the population is homogeneous in that it is almost wholly Muslim. In fact, Indonesia possesses the world's largest Muslim population.

Once again, this southernmost member of the Southeast Asian sector of the West Pacific Rim is still largely rural (74 per cent of the total population) and agriculture contributes about 25 per cent of GDP, mostly through smallholdings. The country is self-sufficient in rice, but the major plantation crops, rubber and sugar, are the ones that earn so

much of Indonesia's foreign exchange. As for industrialisation, manufacturing contributes over 15 per cent of GDP and much of it is in the
hands of foreign investors and companies from Japan, Taiwan, South
Korea and Hong Kong, from where many companies have shifted their
operations to lower-cost locations like Indonesia.

All the indicators suggest that Indonesia's economy is pressing ahead
successfully, and it has been claimed that the Fifth Five-Year Plan
(1989–93) could involve Indonesia in an upwards 'spiral of deregulation,
investment, exports and growth', in which GDP can expand by 5 per
cent a year. Economic reform plans include reducing the present dependence on oil and reducing trade barriers to encourage Indonesia's non-
oil exports, particularly of manufactured products. Without success in
this field there is the growing spectre of unemployment – already
around 10 per cent – with all that means for political as well as economic
stability. There is also growing concern about the glaring income disparities, far greater than in most other countries of the West Pacific Rim
(Figure 1.1). But much will also depend on how Suharto's military-
backed government, with little real political opposition, deals with a
number of political problems on the country's periphery, notably Timor
(to be discussed in Chapter 11) and Irian Jaya. It is only quite recently in
1989 that Indonesia emerged from international hibernation and re-
established its diplomatic links with China, and offered an amnesty to
its remaining ethnic Chinese. Indonesia also has to watch international
opinion over its conservation policy. As one commentator has put it, the
logging industry is 'a motley group of military and Chinese business
interests, who are seen to make huge profits, avoid tax and do little to
replant the forest' (Murray-Brown, 1990, p. 85). Indonesia, far more
than Malaysia, suffers from endemic corruption, inefficiency and a
deeply conservative attitude to many aspects of modern economic life
and thought. Perhaps there can be little real improvement until its
military tradition, which has already lasted twenty-five years under
Suharto, is brought to an end.

Conclusions

Several points can be made about the five countries of the Southeast
Asian sector of the West Pacific Rim. Easily the poorest country,
Vietnam, still has a Communist government running a centrally controlled economy; but, like China, Vietnam is beginning to adopt market-
oriented policies and more open trade policies as a means of breaking
through to greater economic prosperity. All the countries in the region
have a European colonial past and all to varying extents are still primary
commodity-producing countries for a narrow range of minerals and

agricultural products: oil, natural gas, tin, copper, coconut, rubber and palm oil. In this sense they possess colonial-type economies, exporting primary products and until recently importing all the manufactured goods they needed. Now, however, all four countries, and especially Malaysia, have gone some away along the path of industrialisation and are beginning to emphasise non-primary product exports, mostly of low bulk and high value. According to most commentators, all five countries have the potential – for they certainly have the resources – to become NICs in the near future. But, even more clearly than in the countries in the north of the West Pacific Rim, future economic development will depend very much on political factors.

11

PACIFIC CONCERNS

The recent economic success of so many West Pacific Rim countries has depended critically upon their increasingly outward-looking orientation. This eastern maritime fringe of Asia has always been involved in outside contacts, both before and since European penetration. There have been periods of introversion. Japan was was for two centuries largely closed to the outside world before the Meiji Restoration, an event in which the United States played a key role. And China, the nearest to being a continental state in the region, was effectively closed from 1949 until late 1978 when the Reforms initiated the country's greater openness. Today, however, the countries of the West Pacific Rim are beginning to think of themselves as members of the wider Pacific world, including the Ocean's eastern rim, and, increasingly, of the global context in which they must live. This change has resulted above all from two principles pursued in the Pacific since the end of the Second World War: the economic principle of increasing international trade, particularly liberal free trade; and the interrelated security principle of deterrence.

The role of the United States

Since 1945 both these principles have been pursued energetically by the United States. Interpreting the confrontation between the two superpowers, itself and the Soviet Union, very largely in terms of the logic of NATO, the United States took up the interests of the West Pacific Rim, especially Japan. Japan thus became dependent on the United States both economically, for its post-war rebuilding, aid, and trading opportunities, and defensively, for its own security against any threat from the Asian mainland. The logic of the United States's argument was clear: build up a strong economy in Japan through aid and a policy of liberal international trade as a means of deterring the Soviet Union from spreading its power and influence into the Pacific.

The role played by the United States is certainly critical to any understanding of the present prosperity of East Asia. Whereas a good deal of this book has been concerned with examining a wide range of such internal factors as resources, culture, the historical legacy and domestic policies, it may well be true, as Nestor (1990) argues, that it is to the role played by the United States in the region since the Second World War that one must look for a full explanation of the region's dramatic economic growth. While internal factors have played an important permissive or facilitating role, it was the fact that the United States regarded the countries of the West Pacific Rim as vital to its own geopolitical interests that provided the catalyst and the motive power for rapid economic advance in the region. According to Nestor,

> the US nurturing of Japan back to economic health . . . its sponsorship of Japan's return to prominence first in East Asia, then the world economy, are the key reasons for Japan's present economic super-power status. With the United States providing open international markets and military security, Japan's ruling corporatist triad of conservative party, bureaucracy and big business were able to concentrate on leading the economy to ever higher levels of export volumes and technical sophistication through rational industrial and neo-mercantilist trade policies. (Nestor, 1990, p. xi)

Regarding the maritime fringe of East Asia as its first line of defence against the threat of communism and any incipient resurgent imperialism, the United States' aid to and support for the Japanese economy in particular was firm and uncompromising. United States aid and its open-trade policies have been the main instrument of its defence strategy: 'Japan's return to economic hegemony over East Asia was a central pillar of American policy' (Nestor, 1990, p. x). To make this policy work the US encouraged triangular trade between the United States (providing capital and technology), Japan (providing mass consumer and capital goods) and Southeast Asia (providing raw materials). Furthermore, the British in Southeast Asia, working through the Colombo Plan, were encouraged to facilitate this triangular trade policy. This effectively knitted together the economies of the whole of the West Pacific region and turned it into one of the powerhouses of the world economy with growth rates twice those of other regions. The benefits of the rapid growth of the Japanese economy are now being diffused through the NICs to China and to the countries of Southeast Asia (Nestor, 1990, pp. x–xi).

The demise of the former Soviet Union

In 1992 the scene has changed suddenly and dramatically from what it was only a year or two ago. The demise of the Soviet Union leaves the United States as the only super power, and the threat from the former Soviet Union is considered to have largely receded. While looking with some caution at China, the last great communist state, the United States now considers the principle of defence through deterrence to have declined in significance.

Whether or not it is wise to consider that the threat of communism is now greatly reduced, the collapse of the former Soviet Union certainly shifts the focus to Communist China – by far the largest country in the West Pacific Rim and containing perhaps 20 per cent of the world's population. As yet there seems to be no reason why the removal of the threat of communism from the former Soviet Union should now be replaced by fear of a similar threat from China. At least in the economic sphere, both China and its smaller Communist neighbours Vietnam and North Korea now seem to be adopting a largely pragmatic stance in economic policy, including foreign trade and investment. The questions that remain are whether China's present government will survive; whether it will retain its ideological purity and absolute control in the political sphere; and whether such control will continue to inhibit the full implementation of the economic reforms the country so sorely needs.

Nevertheless, the immediate threat from the former Soviet Union territory has now vanished, leaving the United States to consider how far it should continue to play the same supporting economic role in the West Pacific Rim now that the central reason for that role no longer exists. What the United States decides to do has important practical and psychological implications for the entire region. Foremost among these is the growing power of Japan within East Asia and, increasingly, in the wider world economy.

Japanese hegemony

The rise of Japanese economic hegemony has already given rise to a number of changes. While the triangular trade referred to earlier has continued, it now does so within a new division of labour in the region. The United States and Japan now compete as industrial powers, both supplying capital goods and finance; Japan's mass consumer products are now challenged, not by the United States but by the four NICs and China; and the Newly Exporting Countries (NECs) of Malaysia, Indonesia and the Philippines, together with Thailand, now form the bottom league of the system, supplying raw materials while at the same time building up their own export-oriented consumer industries (Nestor, 1990, p. xi).

Other changes arise from the increasingly dominant role being played by Japan in world affairs, both economic and political. Economically it seems inevitable that the balance of power will shift if, as seems probable, Japan eventually overtakes the United States as the world's largest economy. Politically, however, the problem is very different in that Japan seems prepared only slowly and reluctantly to play an effective international role. This was inevitable. Japan has for several decades depended on the United States for its security in East Asia: Japan has never before been required or expected to take any significant part in international affairs. Only very recently has it been argued that Japan should now be ready to accept the global political and strategic responsibilities which stem from its formidable economic power.

Present concerns

The possible reduction in the United States' strategic and economic commitment to the West Pacific Rim is regarded with ambivalence by most, if not all, countries in the region; this is particularly so in the context of disagreements over trade policy between mercantilist, or neo-mercantilist Japan and liberal free-trade America. As pointed out in an earlier chapter, the dangers of three great world trading blocs increasingly turning in upon themselves are openly recognised and feared. But the countries of the West Pacific Rim also have a number of more specific concerns relating to the demise of the Soviet Union, the growing power of China, and a host of more local problems, any one of which could so easily disturb the political equilibrium on which, as we have seen in a number of cases in these pages, so much of their economic growth and prosperity must depend.

Japan's concerns include the effect the withdrawal of the United States naval, air and military bases from the region could have on the security of Japan's trade routes through the inland seas and straits of the West Pacific Rim. The Straits of Malacca between Malaysia-Singapore and the Indonesian island of Sumatra plays a key strategic role in this respect: 80 per cent of Japan's oil and about one-third of its foreign trade flows through these straits. Moreover, the United States' bases in the Philippines – Clark Air Base and Subic Bay Naval Base – have been critical in the defence of Japan's shipping routes through the South China Sea. Here the strategically important Spratly and Paracel island groups lie across the main trade routes. These are both potential flashpoints because they are claimed by several countries – Spratly by six and

Paracel by two – all of them interested in the rich mineral and fishing resources. The resolution of the disputes over these island groups still awaits ratification of the Law of the Sea Convention. Japan is also concerned over the possible reunification of Korea, for a united Korea could eventually pose a serious military as well as economic threat to Japan, especially if present negotiations over North Korea's nuclear capacity are not successful.

It might be thought that a common problem in creating a sense of nationhood would be the physical fragmentation of several of the countries, including Japan, into many islands. Certainly this factor would seem to operate against rapid integration of any national system, especially in Indonesia and the Philippines; but for the most part this does not seem to be a major cause of centrifugal tendencies within any of these states. Much more significant is the degree of cultural, linguistic, ethnic and religious heterogeneity. Dialect diversity in China and the problem of its minorities, such as the non-Han, non-Chinese speaking Muslims of Xianjiang; and the religious conflict between Catholics and Muslims in the Philippines: these are two outstanding examples of countries affected in this way. But perhaps the clearest example is in Malaysia, where a sense of national unity and full economic integration is peculiarly difficult to achieve, not primarily because of the physical division of the country into three units – two of them in the island of Borneo – but mainly because of the ethnic divisions between the Malays, Chinese and Indians, divisions which are to a significant extent expressed politically and occupationally, as well as geographically.

Some of the current areas of disagreement and potential flashpoints in the West Pacific Rim can be directly attributed to the legacy of the colonial period. Indonesia, for instance, objected strongly to Sarawak and Sabah becoming part of the Malaysian Federation; and the Philippines has occasionally expressed its claims over Sabah. But the two major problems with clear colonial origins which are likely to attract increasing attention are East Timor and Hong Kong.

The former Portuguese territory of East Timor, annexed by Indonesia in 1975, has on a number of recent occasions attracted the attention of the world because of Indonesia's brutal repression of the Timorese. Indonesia wishes to incorporate East Timor as a Catholic province within the world's largest Muslim nation. Portugal and Australia, which has now reconsidered its recognition of the Indonesian annexation, have drawn attention to the plight of East Timor. Located as it is within the vast island world of Indonesia, independence for the small territory is obviously fraught with difficulties. But it is possible that the United Nations Security Council will agree to a referendum in the territory. Such a plebiscite would ascertain clearly whether East Timor wishes to remain part of Indonesia, to revive its links with Portugal or to become independent. As with Hong Kong, the immediate future of East Timor is being viewed as a matter of global significance.

Hong Kong is perhaps the most critical case in any consideration of the geopolitics of the West Pacific Rim. It is still a colony but is to be taken back into the control of mainland China in 1997. It lies at the interface between the communist and non-communist world, it is itself a showpiece of what free trade and market capitalism can achieve, and it will become the touchstone of how far China is prepared or able to enter more fully into the global economy: without this, Hong Kong's future economic prosperity is uncertain. Indeed, China's strategic and economic role at both regional and global levels will be judged by how far it fulfils its agreed obligations to the territory of Hong Kong. However, as suggested in Chapter 8, there are reasons for some optimism on the grounds that economic imperatives will ensure that Hong Kong continues to play its present critical economic role and, indeed, that it could well act as a catalyst for China's economy, at least in the southeastern part of the country.

Other regional problems go back to the end of the Second World War in 1945. On that date the Soviet Union occupied four islands in the Kuriles and Japan has consistently argued for their return. Now that Japan has to deal with a quite different, weaker and less powerful government in Russia it is at least possible that this problem will be resolved peacefully in Japan's favour. The conflict between North and South Korea has its origins in the determination of the West, through the United Nations, to prevent the former Japanese colony being overrun by Chinese Communism. Now, however, the collapse of its economy has forced North Korea to develop trading and cultural links with South Korea. As for Taiwan, its political separation from mainland China similarly reflects the ideological struggle between Western democracy and Chinese Communism; but it no longer seems to be a serious flashpoint in the region since the new Kuomintang Party in Taiwan has decided to cling no longer to the fiction of regaining control of mainland China.

Conclusions

There seems little doubt that the major underlying concern of West Pacific Rim countries lies in the possible removal, or at least a significant reduction, of the United States presence in and commitment to their region. This would be likely to increase political instability and tension. It is recognised that none of the present regional arrangements, including ASEAN, could be expected to provide the security umbrella previously provided by the United States. This is not to suggest that the

United States plans a complete withdrawal. But the relaxed manner with which the United States agreed to remove its bases from the Philippines has disturbed some observers, though in reaching a recent 'memorandum of understanding' between the United States and Singapore it was stated by the United States that such a memorandum is 'a good example of the kind of new arrangements our military would like to develop in Southeast Asia. We will be pursuing these kinds of arrangements with a number of other countries – our military is prepared to distribute its presence in the region' (The Times, 1992, p. 6).

The desire to keep the United States engaged in the region also derives from a strong fear of Japanese economic hegemony in the West Pacific Rim countries. Many countries fear that they are now on the threshold of an economic Pax Nipponica, in which the growing regional and global power of Japan will increasingly determine the nature and extent of their own economic growth. There is also some concern being expressed at Japan's heavy military expenditure, some evidence of a growing nationalism in the country, and the possibility of a resurgence of Japan's pre-war militarist, imperialist ambitions. However unwarranted, such fears are beginning to be expressed, not only in the West Pacific Rim but also further south in Australasia. What is called the 'benign superpower', the United States, will, it is believed, provide a necessary 'buffering presence' in the varied and complex world of the West Pacific Rim. Furthermore, the continued engagement of the United States in the region is likely to reduce the possibility of trade blocs developing. East Asian countries are already concerned that the deeper integration of the European Community into a 'Fortress Europe' and the rise of the North American Free Trade Zone (Canada, United States and Mexico) could mean the eventual creation of separate trade blocs. For the West Pacific Rim, this would mean setting up defensive measures such as an East Asia Economic Caucus, a kind of resurrected Greater East Asia Co-Prosperity Sphere. This could be extended to include the Australasian world.

Yet this would go against the interests of the region as they are perceived today. All the countries of the West Pacific Rim have made notable economic progress in recent years. This progress has been built on the opportunity to participate in international trade. If these trading opportunities are removed then the economic growth of all states in the region must suffer. 'The East Asian and Southeast Asian countries do not ask for much assistance from the North, but rather demand that they open their markets wider to exports from East Asia, increase investment in the region, and speed up technology transfer' (Dickson and Harding, 1987, p. 27). While the United States remains fully engaged in the West Pacific Rim, it is believed that these demands have at least a chance of being met.

Conclusion

The West Pacific Rim is an area of great diversity, containing within it a bewildering confusion of environments, cultures, languages, religions and economic systems. It also includes the world's most successful economy, the last great communist state and the most populous Islamic country. Levels of economic development range from low-income to high-income economies, but a survey of the economies in the region indicates that the area as a whole has been relatively very successful in its development over the past ten to twenty years. In particular, growth rates have been high and sustained, even in the poorer countries like China. Economically this is the fastest growing area in the world, and 'the rise of East Asia to a position of wealth and power is one of the major forces shaping the international economic and political system in the latter half of the twentieth century' (Perkins, 1989, p. 3). Putting this dramatic economic surge into its historical context, Figure C.1 illustrates how rapidly some of the West Pacific Rim countries have improved their productivity compared with the older Western economies.

In trying to explain this success – the main purpose of this book – there seems at first sight very little evidence to support any kind of geographical, historical or cultural determinism. However, this conclusion does not preclude the possibility that certain factors have been especially helpful, permissive or facilitating. This seems to be particularly true of four factors.

First, the insular, peninsular and littoral nature and the geographical position of the West Pacific Rim have throughout history encouraged contacts both between the various component states within the region and from outside. This has not only encouraged complex trading networks, it has also fostered a more outward-looking and less parochial attitude among its peoples than was possible in great continental areas like sub-Saharan Africa. Furthermore, with the rise of continental communism in Asia, the West Pacific Rim came to be viewed geopolitically, especially by the United States, as a critical 'line of defence' against the

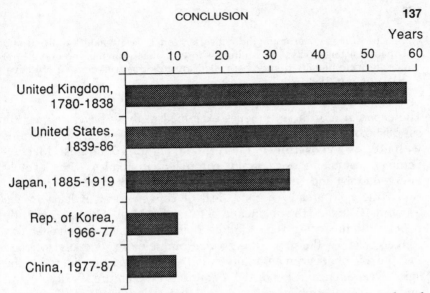

Figure C.1 Periods during which output per person doubled in selected countries (Source: World Bank, 1991)

outward spread of communism. Secondly, while historical factors cannot be cited as an explanation for the regions's success – colonialism, for instance, seems only ever to be used to explain failure – the more recent history of the West Pacific Rim countries since the end of World War II in 1945 has certainly helped economic growth: enlightened self-interest led to the critical period of United States' assistance, especially for Japan and her former colonies, in terms of aid and trade. Thirdly, while serious doubts have been cast on the role of culture as a determinant of economic growth within particular societies or economies, it seems that the Chinese diaspora is a particular case of cultural diffusion or culture contact which has operated as a catalyst for development in many of the host countries.

Finally, government intervention seems to have played a critical role in the region's economic success. It is sometimes suggested that the amount of government intervention is great in the command economies and only slight in the 'free-market' economies. In fact, however, there is little correlation between the amount of government intervention and the extent to which an economy is or is not 'free-market'. As Wade (1990) has shown so convincingly, what matters is not the *amount* of government intervention but the *way* allocation decisions are divided between markets and public administration. Governments may also be rigid and doctrinaire or flexible and selective in their intervention. In the 'miracle' economy of South Korea, for instance,

the government intervened freely and actively in the running of the economy

. . . the Korean government did not suffer from the 'soft-state' syndrome and it had a strong political commitment to economic growth. The energies of society were mobilised and channelled towards this well-defined objective. (Das, 1992, p. 190).

The experience of the economies examined in these pages refutes the simplistic opposition of (i) capitalist, free-market, democratic countries with little or no government intervention, and (ii) communist, Marxist-Leninist, command or centrally controlled economies with severely interventionist and repressive governments. As noted on several occasions, there is already a good deal of convergence, at least in the economic field; and the distinction between the ideal of liberal free-trade policies and the reality of mercantilist policies seems likely to hasten this convergence. Furthermore, it is now becoming more generally accepted that the role of government intervention needs to be reconsidered in the light of experience: it is not the amount but the nature, flexibility and purpose of government intervention that is important. Government intervention is critical in co ordinating business effort and in encouraging risk-taking.

Suggested answers to the questions

In attempting to answer the three groups of questions posed in the Introduction it is impossible to give any firm, unequivocal answers. All one can do is to put forward one's own opinions, based on the kinds of material and arguments presented in these pages and based, too, on one's own necessarily biased and selective experience in some countries of the West Pacific Rim.

Recalling the first of the three groups of questions – those concerning explanations of the region's remarkable success over the past decade or so – it will already be clear that, in the writer's opinion, it seems difficult to avoid the conclusion that the engine of economic growth in the West Pacific Rim countries has been and remains trade. Where there has been little trade there has been little growth. For instance, it was argued in Chapter 5 that China's refusal to allow domestic trade to develop before 1978, and its determination to control all exchange through its own state system of procurement and distribution seriously inhibited internal growth and still constrains China's operation of its present 'open-door' policy. In Japan and the NICs, on the other hand, it is inconceivable that such rapid economic growth would have occurred without the energetic trade policies consistently pursued by their governments. It is now generally accepted in all the region's territories, including the communist countries, that trade, whether domestic or international, is critical not only in making possible the development of a modernised industrial economy, but also in fostering the necessary outward-looking mental attitudes and ways of thinking.

Regarding the second group of questions, about whether the experience of the West Pacific Rim countries helps to throw new light on approaches or attitudes to the study of economic growth, it is clear from these pages that 'wealth creation' or 'growth' strategies have proved the most successful, and that 'growth with equity' policies, emphasising the need for wealth redistribution, both geographically and structurally, are now being abandoned, even in the two communist countries. Political and social imperatives everywhere demand rapid, sustained and perceptible aggregate economic growth. Even China now accepts the need to pursue a policy which includes the encouragement of 'growth poles' and their so-called 'trickle-down' effects. The policies embraced so enthusiastically by China in 1949 and supported by many Western academics have failed. Indeed, 'there can be no more powerful a demonstration of the fact that Marxism has been the great philosophical and economic failure of the century as a model for capital formation and the raising of living standards than in the West Pacific Rim' (Winchester, 1991, p. 73).

It is now possible to identify and confront a number of assumptions about economic growth in the West Pacific Rim. Most important is the implicit – and occasionally explicit – search for a set of determining factors which some believe to be universal, which some believe apply to a particular group of countries, and which others believe are unique and specific to each country. It is assumed that once this set, or combination, of factors is identified and its course accurately charted, then it should be possible to construct a model, or at least to imitate the set of circumstances and to map a similar course to bring about successful economic growth.

It is partly for this reason that discussion about economic growth in East Asia so often implies a belief in the need to make simple, clear choices between sets of crude alternatives: communism or capitalism; democracy or authoritarianism; government intervention or the free market; the formal and informal sector; urban or rural bias; agriculture or industry; ISI or EOI; and top-down or bottom-up strategies. The list is endless. In practice, however, the distinctions are far from clear. Reality is usually an amalgam which provides what one might call a country's unique cultural signature: the way in which a people organise their economic, social and political life, and how they deal with the problems that arise. This cultural signature is revealed in their aspirations, philosophies and attitudes; it involves their reactions to established views and forms or organisation, to new ideas and exotic stimuli; and it is in part a consequence of

environment, history and chance events. Economic growth throughout the countries of the West Pacific Rim has been time- and place-
specific; and each country, encouraged and sustained no doubt by
the examples of its most successful neighbours, will continue to pursue its economic growth against the background of what we have
called its unique cultural signature.

Arguably there has been too much emphasis on the identification and
explanation of what are believed to be determining, or causal, factors.
Observers have become entangled with the philosophy of developmentalism. But this is no more than a confusion of the moral values, pseudo-
science and self-righteous proselytising which characterised the early
missionaries in the region; more often than not it is based on distorted
and half-understood interpretations of other societies and other values.
Consequently there has been a tendency to lose sight of what may be the
simple truth: that the central concern should be whether or not the
unique characteristics and evolutionary path followed by an economy
prevents, permits or facilitates the extension and institutionalisation of
trade. There is no 'correct' set of institutional or policy formulae, nor is
there any 'correct' set of determining factors that will enable a country to
unlock its economic growth. Whatever is expedient in encouraging or
facilitating the extension and institutionalisation of trade will bring
about economic growth.

As for the third group of questions about the global political and
strategic role of the West Pacific Rim countries, the answers are today
very different from what they would have been only a year or two ago.
The disintegration of the former Soviet Union and the demise of
Communism there means that the perceived threat to the West and to
the Pacific Basin countries has now receded very considerably, if not
entirely. It is true that China, with its vast population and resources
remains a tightly controlled communist state. But the consensus of
opinion seems to be that the threat from China is economic rather than
strategic or military, and the previous chapter has already referred to the
possible changes in the balance of power, both in the Pacific and
globally, should China develop into a major trading nation early in the
next century.

Moreover, there are already signs that China could increasingly play a
role as regional peacekeeper: it has already helped in improving relations between North and South Korea, and in preserving the balance of
power in Indonesia. The growth of trans-Pacific trade between the West
Pacific Rim countries, including China, and the East Pacific Rim
countries has already led to important intellectual cultural and political
contacts across the Pacific; and these will no doubt develop still further
unless the United States decides to reduce its presence in the region.
Above all, however, these pages suggest that continued economic

growth in the West Pacific Rim will confirm and increase the region's global role; and that Japan's economic power will challenge and may well replace the present fragile hegemony of the United States during what some like to call the Pacific Century.

References and further reading

Aikman, D. (1986) *Pacific Rim: area of change, area of opportunity*, Boston, Little, Brown.

Alip, E.M. (1974) *The Chinese in Manila*, Manila, National Historical Commission.

Allen, G.C. (1981) *The Japanese Economy*, London, Weidenfeld and Nicolson.

Anderson, K. and Y. Hayemi (1988) *The Political Economy of Agricultural Protection: East Asia in comparative perspective*, London, Allen & Unwin.

Andors, S.P. (1984) China's industrialization in historical perspective, in N. Maxwell (ed), *China's Road to Development*, Oxford, Pergamon.

Ariff, M. and L.H. Tan (1988) *ASEAN Trade Policy Options*, Singapore, ISEAS.

Awanohara, S. (ed.) (1990) *Japan's Growing External Assets: a medium for regional growth*, Hong Kong, OUP.

Balassa, B. (1988) The lessons of East Asian development: an overview, *Economic Development and Cultural Change (EDCC)*, supp. 36, m. 3 (April) pp. S273–90.

Bannister, J. (1987) *China's Changing Population*, Stanford University Press.

Barker, R. and R.W. Herdt with Beth Rose (1985) *The Rice Economy of Asia*, Washington DC, Resources for the Future.

Barrett, R.E. and S. Chin (1987) Export-oriented industrialising states in the capitalist world system: similarities and differences, in Deyo (1987) pp. 23–43.

Barrett, R. and M.K. Whyte (1982) Dependency theory and Taiwan: a deviant case analysis, *American Journal of Sociology*, vol. 87, pp. 1064–89.

Bauer, P.T. (1981) *Equality, the Third World and Economic Delusion*, London, Weidenfeld & Nicolson.

Bauer, P.T. and B.S. Yamey (1957) *The Economics of Underdeveloped Countries*, Cambridge University Press.

Beasley, W.G. (1990) *The Rise of Modern Japan*, London, Weidenfeld and Nicolson.

Benewick. R. and P. Wingrove (eds) (1988) *Reforming the Revolution: China in transition*, Basingstoke, Macmillan.

Beresford, M. (1988) *Vietnam*, London, Pinter.

——. (1989) *National Unification and Economic Development in Vietnam*, London, Macmillan.

Berger, P.L. and Hsiao, H.H.H. (eds) (1988) *In Search of an East Asian Development Model*, New Brunswick, Transactions.

Berger, R. (1975) Economic planning in the People's Republic of China, *World Development*, vol. 33, pp. 552–4.

Berstein, T.P. (1977) *Up to the Mountains and Down to the Villages: The Transfer of Youth from Urban to Rural China*, Yale University Press.

Bishop, C.W. (1922) The geographical factor in the development of Chinese civilisation, *Geographical Review*, vol. 12, pp. 19–41
——. (1932) The rise of civilisation in China with reference to its geographical aspects, *Geographical Review*, vol. 23, pp. 617–31
Blaker, J. (1970) *The Chinese in the Philippines*, Ann Arbor, Mich., University of Michigan.
Blecher, M. (1986) *China: politics, economics and society*, London, Pinter.
Boeke, J.H. (1966) *Indonesian Economics: the concept of dualism in theory and practice*, The Hague, Van Hoeve.
Bristow, R. (1984) *Land-Use Planning in Hong Kong*, Hong Kong, OUP.
Bradford, C.I. and W.H. Branson (eds) (1987) *Trade and Structural Change in Pacific Asia*, University of Chicago Press.
Brugger, B. (ed) (1985) *Chinese Marxism in Flux, 1978–1984*, London, Croom Helm.
Buchanan, K. (1966) *The Chinese People and the Chinese Earth*, London, Bell.
——. (1970) *The Transformation of the Chinese Earth*, London, Bell.
Buck, D.D. (1978) *Urban Change in China – Politics and Development in Tsinan, Shantung, 1890–1949*, Madison, Wisc., University of Wisconsin.
——. (1981) Policies favouring the growth of smaller urban places in the People's Republic of China, 1949–1979, in Ma and Hanten (1989) pp. 73–91.
——. (1987) The study of urban history in the People's Republic of China, *Urban History Yearbook, 1987*, Leicester, pp. 61–75.
Buck, J.L. (1937) *Land Utilisation in China*, Shanghai, University of Shanghai.
Bucknall, K. (1989) *China and the Open Door Policy*, Sydney, Allen & Unwin.
Cannon, T. and A. Jenkins (eds) (1990) *The Geography of Contemporary China: the impact of Deng Xiaoping's decade*, London, Routledge.
Castells, M., G. Lee and R. Yin-Wang Kwok (1990) *The Shek Kip Mei Syndrome: economic development and public housing in Hong Kong and Singapore*, Hong Kong, Pion.
Cheng, Lim-Keak (1985) *Social Change and the Chinese in Singapore: a socio-economic geography with specific reference to bang structure*, Singapore University Press.
Chiu, H., Y.C. Jao and Y.L. Wu (eds) (1987) *The Future of Hong Kong: towards 1997 and beyond*, New York, Quorum.
Cole, J. (1987) *Development and Underdevelopment: a profile of the Third World*, London, Methuen.
Coppel, C.A. (1983) *Indonesian Chinese in Crisis*, Kuala Lumpur, OUP.
Copper, J.F. (1990) *Taiwan: Nation-State or Province*, Boulder, Col., Westview.
Croll, E., D. Davin and P. Kane (1985) *China's One-Child Family Policy*, London, Macmillan.
Cumings, B. (1984) The origins and development of the NE Asian political economy: industrial sectors, product cycles and political consequences, *Industrial Organisation*, vol. 38, no. 1 (Winter) pp. 11–40.
Cushman, J. and Wang Gungwu (1988) *Changing Identities of the Southeast Asian Chinese since World War II*, Hong Kong, HK University Press.
Das, D.K. (1992) *Korean Economic Dynamism*, Basingstoke, Macmillan.
Deyo, F.C. (ed) (1987) *The Political Economy of the New Asian Industrialism*, Cornell University Press.
Dickson, B. and H. Harding (eds) (1987) *Economic Relations in the Asia-Pacific Region*, Washington, DC, Brookings Institute.
Dixon, C. (1991) *South East Asia in the World-Economy*, Cambridge University Press.
Donnithorne, A. (1967) *China's Economic System*, London, Hurst.
Dore, R. (1986) *Flexible Rigidities: industrial policy and structural adjustment in the*

Japanese economy, Stanford University Press.

Downton, E. (1986) *Pacific Challenge*, Toronto, Stoddart.

Drakakis-Smith, D. (1992) *Pacific Asia*, London, Routledge.

Drake, C. (1989) *National Integration in Indonesia*, Honolulu, University of Hawaii Press.

Drysdale, P. (1988) *International Economic Pluralism: economic policy in East Asia and the Pacific*, Sydney, Allen & Unwin.

Dutt, A.K. (ed) (1985) *Southeast Asia: realm of contrasts*, Boulder, Col., Westview.

Dwyer, D. (ed) (1990) *Southeast Asia: geographical perspectives*, Harlow, Longmans.

East, W.G., O.H.K. Spate and C.A. Fisher (eds) (1971) *The Changing Map of Asia*, London, Methuen.

Economist, The (1991a) 5 October, p. 14.

——. (1991b) Asia's emerging economies, 16 November, pp. 5–24.

——. (1991c) *The World in 1991*, London, The Economist.

Edmonds, R.L. (1991) China's environment: problems and prospects, unpublished paper, University of Keele, pp. 1–31.

Elster, J. and K. Moene (eds) (1989) *Alternatives to Capitalism*, Cambridge University Press.

Far Eastern Economic Review (FEER) (1988), Sizzling hot chips: Asia is the source of the semiconductor industry's spectacular growth, 18 August, pp. 80–86.

Far Eastern Economic Review (FEER) (1990a), 15 March, pp. 38–9.

Far Eastern Economic Review (FEER) (1990b), 31 May, pp. 57–8.

Fisher, C.A. (1971) 'South East Asia' and 'The maritime fringe of East Asia: Japan, Korea and Taiwan,' in East, Spate and Fisher (1971) pp. 221–339; 449–560.

Fitzgerald, C.P. (1966) *A Concise History of East Asia*, London, Penguin.

Geertz, C. (1963) *Peddlers and Princes: social change and economic modernisation in two Indonesian towns*, University of Chicago Press.

Gereffi, G. (1989) Development strategies and the global factory, in Gourevitch (1989) pp. 92–104.

Gereffi, G. and D. Wyman (eds) (1990) *Manufactured Miracles: development strategies in Latin America and East Asia*, Princeton University Press.

Gerrard, B. (1989) *A Theory of the Capitalist Economy: towards a post-classical synthesis*, Oxford, Blackwell.

Gerschenkron, A. (1962) *Economic Backwardness in Historical Perspective*, Harvard University Press.

Gibson, L.J. (1990) The Pacific Rim: region or regions, *Papers of the Regional Science Association*, vol. 68, pp. 1–8.

Gold, T. (1986) *State and Society in Taiwan's Economic Miracle*, NY, Armank.

Goodman, D.S.G. (1988) *Communism and Reform in East Asia*, London, Frank Cass.

——. (ed) (1989) *China's Regional Development*, RIIA, London, Routledge.

Gourevitch, P.A. (ed) (1989) The Pacific Region: challenges to policy and theory, Special Edition, *Annals of the American Academy of Political and Social Science (AAPSS)*, vol. 505 (September).

Guzman, R.P. de and M.A. Reforma (1988) *Government and Politics of the Philippines*, Oxford University Press.

Hall, K.R. (1985) *Maritime Trade and State Development in Early Southeast Asia*, Honolulu, University of Hawaii.

Hamilton, G.G. and N. Woolsey Biggart (1988) Market, culture and authority: a comparative analysis of management and organisation in the Far East, *American Journal of Sociology*, supp. 94, pp. S69–74.

Hao Yen-p'ing (1986) *The Commercial Revolution in Nineteenth Century China: the rise of Sino-Western mercantile capitalism*, Berkeley, Calif., University of California.

Hebbert, M. and Norihiro Nakai (1988) How Tokyo grows: land development and planning on the metropolitan fringe, Suntory Toyota International Centre for Economics and Related Disciplines (ST/ICERD), *ST/ICERD*, occasional paper, no. 11, London, ST/ICERD.

Heng Pek Koon (1988) *Chinese Politics in Malaysia: a history of the Malayan Chinese Association*, Singapore, OUP.

Higgott, R. and R. Robison (eds) (1985) *Southeast Asia: essays in the political economy of structural change*, London, Routledge & Kegan Paul.

Hilbert, D. (1988) Closing remarks: about economic growth – a variation on a theme, *EDCC*, supp. 36 no. 3 (April) pp. S291–307.

Ho, Ping-ti (1966) *An Historical Survey of Landsmannschaften in China*, Taipei, Govt. of Taiwan.

Hoare, J. and S. Pares (1988) *Korea: an introduction*, London, Kegan Paul International.

Hodder, R.N.W. (1990) China's industry – horizontal linkages in Shanghai, *Transactions, Institute of British Geographers*, New Series, vol. 15, pp. 487–503.

——. (1991) Planning for development in Davao City, The Philippines, *Third World Planning Review*, vol. 13, no. 2, pp. 105–28.

Hofheinz, R. (Jr) and K.E. Calder, (1982) *The Eastasia Edge*, New York, Basic Books.

Howe, C. (1978) *China's Economy: a basic guide*, London, Elek.

Howe, C. and K. Walker (1989) *The Foundations of the Chinese Planned Economy: a documentary survey, 1953–65*, London, Macmillan.

Hsiao, M.H.H. (1988) An East Asian development model: empirical approaches, in Berger and Hsin-Huang (eds), 1988, pp. 12–23.

Hughes, H. (ed) (1988) *Achieving Industrialization in East Asia*, Cambridge University Press.

Inoguchi, T. (1989) Shaping and sharing Pacific dynamism, in Gourevitch (1989), pp. 46–55.

Inoguchi, T. and D.I. Okimoto (1988) *The Political Economy of Japan*, Vol. 2, Stanford University Press.

Ishikawa, S. (1967) *Economic Development in Asian Perspective*, Tokyo, Institute of Economic Research, Hitotsubashi University.

Jacobs, N. (1985) *The Korean Road to Modernisation and Development*, Urbana, University of Illinois.

Johnson, D.G. (1988) Economic reforms in the People's Republic of China, *EDCC*, supp. 36, no. 3 (April) pp. S225-S245.

Jones, L. and I.I. Sakong (1980) *Government, Business and Entrepreneurship in Economic Development: the Korean case*, Harvard University Press.

Kahn, H. (1979) *World Economic Development*, Boulder, Col., Westview.

Kajima, R. (1987) *Urbanisation and Urban Problems in China*, Tokyo, Institute of Developing Economies.

Kallgren, J.K. et al (eds) (1988) *ASEAN and China: an Evolving Relationship*, Berkeley, Calif., University of California.

Kolb, A. (1971) *East Asia*, London, Methuen.

Krause, L.B. (1988) Hong Kong and Singapore; twins or kissing cousins, *EDCC*, supp. 36, no. 3 (April) pp. S45–66.

Kuznets, P.W. (1988) An East Asian model of economic development: Japan, Taiwan and South Korea, *EDCC*, supp. 36, no. 3 (April) pp. S11–43.

Kuznets, S. (1955) Economic growth and income inequality, *American Economic*

Review, vol. 45, no. 1, pp. 1–28.

Landsberger, S.R. (1989) *China's Provincial Foreign Trade: Statistics, 1978–88*, London, R11A.

Lau, L.J. (1986) *Models of Development: a comparative study of economic growth in South Korea and Taiwan*, San Francisco, IEAS.

Lee, C.H. and S. Naya (1988) Trade in East Asian development with comparative reference to S.E. Asian experiences, *EDCC*, supp. 36, no. 3 (April) pp. S123–152.

Lee Yong Leng (1982) *Southeast Asia: essays in political geography*, Singapore, Institute of Southeast Asian Studies.

Leeming, F. (1985) *Rural China Today*, London, Longmans.

Leifer, M. (1983) *Indonesia's Foreign Policy*, London, RIIA.

Liang, K.S. and C.I. Hou Liang (1988) Development policy formation and future policy priorities in the Republic of China, *EDCC*, supp. 36, no. 3 (April) pp. S67–101.

Lieberthal, K. and M. Oksenbery (1988) *Policy Making in China: leaders, structures and processes*, Princeton University Press.

Lin, C.Y. (1988) East Asia and Latin America as contrasting models, *EDCC*, supp. 36, no. 3 (April) pp. S153–97.

Lin, J.Y. (1988) The household responsibility system in China's agricultural reform: a theoretical and empirical study, *EDCC*, supp. 36, no. 3 (April) pp. S199–224.

Lin Shuiyuan (1990) China's industrial structure and the present industrial adjustment, Beijing, Institute of World Economies and Politics, Chinese Academy of Social Sciences.

Lincoln, J.R. and K. McBride (1987) Japanese industrial organisation in comparative perspective, *Annual Review of Sociology*, vol. 13, pp. 67–81.

Lippit, V.D. (1987) The economic development of China, (reviewed in *EDCC*, vol. 37, no. 3 (April) pp. 641–5).

Ma, L.J.C. and E.W. Hanten (eds) (1989) *Urban Development in Modern China*, Boulder, Col., Westview.

Mackie, J.A.C. (1988) Economic growth in the ASEAN region: the political underpinnings, in Hughes (1988) pp. 283–326.

Mackinder, H.J. (1942) *Democratic Ideals and Reality*, New York, Holt.

McGee, T.G. (1967) *The Southeast Asian City*, London, Bell.

Minami, R. (1986) *The Economic Development of Japan: a quantitative study*, Basingstoke, Macmillan.

Murray-Brown, J. (1990) Indonesia, *World of Information*, pp. 83–90.

Myers, R.H. (1988) Land and labour in China, *EDCC*, vol. 36, no. 4 (July) pp. 797–806.

Needham, Sir Joseph (1954) *Science and Civilisation in China*, London, Cambridge University Press.

Neher, C.D. (1991) *Southeast Asia in the New International Era*, Boulder, Col., Westview.

Nestor, W.R. (1990) *Japan's Growing Power over East Asia and the World Economy*, Basingstoke, Macmillan.

O'Malley, W.J. (1988) Culture and modernization, in Hughes (1988) pp. 327–43.

Omohundro, J.T. (1981) *Chinese Merchant Families in Ilolio*, Manila, Manila University Press.

Oshima, H.T. (1988) Human resources in East Asia's secular growth, *EDCC*, supp. 36, no. 3 (April) pp. S103–22.

Perkins, D.H. (1989) *Asia's Next Economic Giant?*, Seattle, University of Washington Press.

Pomfret, R. (1990) *Investing in China*, Brighton, Harvester Wheatsheaf.
Prestowitz, C.V. (1988) *Trading places: how we allowed Japan to take the lead*, New York, Basic Books.
Prybyla, J.S. (1987) *Market and Plan under Socialism: the bird in the cage*, Stanford, Hoover Institution Press.
——. (1990) *Reform in China and other Socialist Economies*, Washington, DC,
Purcell, V. (1952) *The Chinese in Malaya*, London, RIIA.
——. (1965) *The Chinese in Southeast Asia*, London, RIIA.
Rabushka, A. (1987) *The New China: comparative economic development in mainland China, Taiwan and Hong Kong*, Boulder, Col., Westview.
Redding, S.G. (1990) *The Spirit of Chinese Capitalism*, Berlin, Walter de Gruyter.
Riedel, J. (1988) Economic development in East Asia: doing what comes naturally, in Hughes (1988) pp. 1–38.
Rigg, J. (1991) *Southeast Asia: Region in Transition*, London, Unwin Hyman.
Riskin, C. (1987) *China's Political Economy*, Oxford, OUP.
Rostow, W.W. (1962) *Stages of Economic Growth*, 2nd edn, Cambridge University Press.
Rozman, G. (1991) *The East Asian Region: Confucian heritage and its modern adaptation*, Princeton University Press.
Ruttan, V.W. (1988) Cultural endowments and economic development: what can we learn from anthropology?, *EDCC*, supp. 36, no. 3 (April) pp. S247–271.
Scalapino, R.A. (1986) *Economic Development in the Asia Pacific Region*, San Francisco, Institute for East Asian Studies.
Segal, G. (1990a) *Rethinking the Pacific*, Oxford, Clarendon.
——. (ed) (1990b) *Chinese Politics and Foreign Policy Reform*, London, Kegan Paul International.
Sit, V.F.S. (ed) (1985) *Chinese Cities: the growth of the metropolis since 1949*, Hong Kong, Oxford University Press.
Skinner, G.W. (1964–65) Marketing and social structure in rural China, *The Journal of Asian Studies*, vol. 24, no. 1, pp. 3–43; vol. 24, no. 2, pp. 195–228; vol. 24, no. 3, pp. 363–99.
——. (1985) Rural marketing in China: repression and revival, *The China Quarterly*, no. 91, pp. 393–413.
Solinger, D. (1984) *Chinese Business under Socialism: the politics of domestic commerce, 1949–1980*, Berkeley, Calif., University of California.
——. (1985) Commercial reform and state control: structural changes in Chinese trading, *Pacific Affairs*, vol. 38, pp. 197–221.
Soon, L.T. and L. Suryadinata (eds) (1988) *Moving into the Pacific Century: the Changing Regional Order in the Asia-Pacific*, Singapore, Heinemann Asia.
Suryadinata, L. (1989) *The Ethnic Chinese in the ASEAN States*, Singapore, ISEAS.
Tang, A.M. and J.S. Worley (eds) (1988) Why does overcrowded, resource-poor East Asia succeed – lessons for the LDCs? *EDCC*, supp. 36, no. 3 (April) pp. S24–29.
The Times (1992), 4 January, p. 6.
Times Higher Education Supplement (THES) (1991) 18 October, p. 10.
Trewartha, G.T. (1965) *Japan: A Physical, Cultural and Regional Geography*, London, Madison.
Van Nieuwenhuijze, C.A.O. (ed) (1984) *Development Regardless of Culture*, Leiden, Brill.
Vo Nhan Tri (1990) *Vietnam's Economic Policy Since 1975*, Singapore, ISEAS.
Vogel, E. (1979) *Japan as Number One: lessons for America*, Cambridge, Harvard University Press.

——. (1990) *One Step Ahead in China: Guangdong under Reform*, Cambridge, Harvard University Press.

——. (1991) *The Four Little Dragons: the spread of industrialization in East Asia*, London, Harvard University Press.

Wade, R. (1989) What can economics learn from East Asian success? in Gourevitch (1989) pp. 68–79.

——. (1990) *Governing the Market: economic theory and the role of government in East Asian Industrialisation*, Princeton University Press.

Wade, R. and G. White (eds) (1984) Developmental states in East Asia: capitalist and socialist, *IDS Bulletin*, no. 15, Brighton.

Weber, Max (1958) *The Protestant Ethic and the Spirit of Capitalism*, New York, Scribner.

White, G. (ed) (1988a) *Developmental States in East Asia*, IDS, Basingstoke, Macmillan.

——. (1988b), The new economic paradigm: towards market socialism, in Benewick and Wingrove (1988) pp. 81–90

White, G. and Wade, R. (1988), Developmental states and markets in East Asia: an introduction, in White (1988a) pp. 1–29.

White, J.W. (1989) Economic development and sociopolitical unrest in nineteenth-century Japan, *EDCC*, vol. 37, no. 2 (January) pp. 231–60.

Wiltshire, R. (1992) Inter-regional personal transfers and structural change: the case of the Kamaishi steelworks, *Transactions, Institute of British Geographers*, New Series, vol. 17, no. 1, pp. 65–79.

Winchester, S. (1991) *The Pacific*, London, Hutchinson.

World Bank (1990) *World Development Report, 1990: Poverty*, Washington.

——. (1991) *World Development Report, 1991: The Challenge of Development*, Washington.

World of Information (1990) *The Asia and Pacific Review*, Edison, N.J., Hunter.

Wright, A.F. (1966) *The Confucian Persuasion*, Stanford University Press.

Xie, Y. and F.J. Costa (1991) The impact of economic reforms on the urban economy of the People's Republic of China, *The Professional Geographer*, vol. 43, no. 3, pp. 318–35.

Yahuda. M.B. (1990) China, in *World of Information* (1990) pp. 53–61.

Yong, C.F. (1987) *Tan Kah-Kee: the making of an overseas Chinese legend*, Singapore, OUP.

Index

ADB (Asian Development Bank) 113
Africa 8, 10–11, 17, 23, 31, 33, 67, 69, 136
Age structure 8–9, 16–17, 93
Agriculture 6, 13, 26–33, 58, 68, 72, 79, 83–5, 90–92, 99, 101–2, 104, 109–11, 118–19, 121, 124, 126, 139
Aid 67, 129–30, 137
Ainu 38
Aquino, C. 53
Arabs 24–25
ASEAN (Association of Southeast Asian Nations) 65, 99, 119–20, 134
Australasia 44, 38, 69, 74, 133, 135
Australoids 38, 96

Baba Chinese 44
Baguios 12
Bandar Seri Begawan 125
Bang 48–9
Batam 97
Beijing 49, 53–4, 60, 106, 109, 114
Benin 8
Birth control: *see* population control
Birth rate 16–17, 29, 96: *see also* birth control, population
Borneo 123, 125, 133
Brazil 114
Britain 25, 45, 47, 70, 94, 97, 123, 130
Buddhism: *see* religion
Bureaucracy 22, 57, 60, 114, 120, 130
Bushido theory 38

Calvinism: *see* religion
Cambodia 1, 119
Canada 44, 69, 72, 96

Cantonese 44, 95, 98
Capitalism 3, 26–8, 32, 46, 54, 58, 79, 81, 103, 112, 115, 119, 125, 134
Cash crops: *see* agriculture
Catholicism: *see* religion
Chaebol 75, 101–4
Chinatowns 51–2
Chinese diaspora 4, 43–55, 137
Christianity: *see* religion
Clan 49, 51
Clark Air Base 132
Class 8, 28, 37–8, 53, 91
Climate 11–12, 21, 28, 91, 100, 104, 110, 119, 121, 123
Colonialism 4, 25–33, 45–6, 65, 70, 96, 118–19, 122, 127–8, 133, 137
Commerce 4, 31, 36, 39, 41, 47, 54, 56, 60, 83, 123
Culture 4, 18, 21, 24, 28, 34–42, 48, 89, 95, 106, 123, 130, 133, 136–7
Cultural revolution 85
Communes 110
Communications: *see* transport
Communism 3, 5, 21–2, 36, 54, 67, 79–80, 92, 98, 100, 102–3, 107–9, 115, 117, 120–2, 127, 130, 134, 136–40
Confucianism: *see* religion
Conservation 66, 104, 127: *see also* pollution
Corruption 52, 61, 107, 127
Cronyism 121

Death rate 16–17, 29
Debt 28, 49, 121
Democracy 3, 35, 91, 96, 101–2, 104, 115, 121, 126

Deng Xiaoping 54
Dependency 3–4, 9, 27, 88
Deterrence 129–31
Deutero-Malays 38
Dialects 44, 48–53, 95, 98, 104, 133
Diet 8, 10, 66
Doi Moi 85, 119
Dualism 39
Dutch 25, 28, 30, 39, 45, 70, 126

EAEC (East Asian Economic Caucus) 135
Economic strategies 5, 79–89, 92
Education 3, 7–9, 14, 18, 20–1, 42, 44, 49, 51–2, 91–3, 96, 120–1, 123, 125–6
EEC (European Economic Community) 72, 74, 97, 135
Elites 21, 27, 52–3
Entrepreneurs 4, 21, 28, 32, 48, 54, 56, 58, 64, 67, 82, 87, 94, 98–9, 120
Environment 11, 14, 22, 30, 37, 93, 104, 106–7, 136, 139
EOI (Export-oriented industrialisation) 79, 83, 86, 94, 101, 104, 106, 109, 122, 124, 139
EPZ (Export Processing Zone) 122
Ethnicity 21, 29–30, 33, 36–7, 44, 50, 53, 95, 123, 127, 133
Europe 1, 4–5, 25–6, 28–30, 39, 45–7, 52, 65, 69, 74, 81, 93, 102, 115, 127, 129, 135
Exchange 56–64: *see also* trade, procurement and distribution

Family 35–7, 39, 48, 51, 53, 57, 62, 91
Family planning: *see* population control
Feudalism 37
Food 13–14, 29–30, 44, 72, 80, 86, 90, 93, 95, 99, 101, 109, 115, 126
Formosa 103
France 25, 28–9, 119
Fujian 45, 75, 106, 110, 113–4
Fukien: *see* Fujian

Ganqing 49
GATT (General Agreement on Tariffs and Trade) 72
Glasnost 81, 115
Government intervention 5, 79–89, 94, 106, 115, 137–8

Guangdong 45, 75, 99, 110, 113–4
Guangxi 45, 49

Hainanese 44, 95
Hakka 44, 95
Hanoi 85
Health 3, 7–8, 17–20, 29, 96, 108, 125–6
Hinduism: *see* religion
History 4, 11, 23–33, 41, 106, 130, 136–9
Hokkaido 11
Hokkien 44, 51, 95
Horizontal linkages 60, 63
Housing 17–18, 95, 98

Iban 123
IMF (International Monetary Fund) 120
Imperialism 23, 67, 130, 135
Income inequality 6, 8–10, 103, 127
India 23–4, 27, 38–9, 55
Indo-China 38
Industry 6, 10, 13, 26–7, 32, 38, 41, 62, 75, 79–80, 83, 86, 91–2, 95, 98, 101–2, 105, 110–12, 122, 124, 127, 139
Institutions 21–2, 26, 36–7, 41–2, 81–2
Investment 4–5, 14, 65–78, 82, 92–5, 97, 99, 108–9, 111–13, 115–17, 119, 122, 124, 127, 131, 135
Irian Jaya 127
Irrigation 29, 32, 90
ISI (Import substitution industrialisation) 79, 83, 86–7, 94, 104, 106, 109, 122, 124, 126, 139
Islam: *see* religion

Jakarta 25
Java 15, 28, 30, 54
Johore 97

Kadazans 123
Kapampangan 53
Kaohsiung 105
Kinship 48
Kowloon 97
Kuala Lumpur 123
Kuching 47
Kuomintang 134
Kuriles 134

Labour 14, 29, 45–6, 66, 76–7, 86, 96–7, 99, 101, 104, 106, 124
Labuan 25
Laissez-faire policies 81, 87–9, 94, 98
Land reform 29, 84, 91, 122
Language 37–8, 44, 49–50, 53, 91, 95, 121, 133, 136
Laos 1
Latin America 4, 8, 10, 23, 33, 67, 69, 88
Leytonias 53
Li Peng 115–16
Liaioning 75, 113
Life expectancy 8–9, 16–17, 108
Linyuan 104
Literacy 21, 91
Living standards 17, 96, 107, 109, 125, 139
Localism 50, 63
Long-distance trade 58
Lui Kuan 49
Luzon 25

Macao 1, 26
Madura 15
Malacca 25
Manchuria 25
Mandarin 95, 104
Manila 24, 39, 52
Manufacturing 13, 26, 30, 70, 94–5, 122
Mao Zedong 10, 35–6, 57–8, 109
Marcos, F. 53, 101
Marcos, I. 53
Market places 57–8: *see also* trade (domestic)
Marxism 7, 79, 88, 109, 131
Mauritania 8
Meiji restoration 21, 37–8, 90–1, 120
Mekong River 118
Mercantilism 77–8, 83, 132, 138
Mestizos 44, 52
Mexico 135
Micronesia 1
Middle East 67, 69
Migration 15–17, 23–4, 28, 33, 38, 46–7, 95–6, 98–9, 104, 123, 125
Mindanao 24, 39, 121
Minerals 12–13, 27, 30, 45–6, 52, 72, 84–6, 100, 107–8, 118, 120–27, 133
Model 3–4, 42, 90, 106, 139
Mongolia 23–4, 38

Multinational corporations 53, 65, 77, 94, 97, 101
Muslims: *see* religion

Nanjing 51
Nantze 105
Nanyang 45
Nationalism 36, 52, 72, 124, 135
NAFTZ (North America Free Trade Zone) 135
NATO (North Atlantic Treaty Organisation) 129
NDP (National Development Policy) 125
NEC (Newly Exporting Country) 131
Negroids 38
NEP (New Economic Policy) 123–4
New Territories 97
NEZ (New Economic Zone) 119
NICs (Newly Industrialised Countries) 1, 3, 5, 8, 65, 67–8, 75, 77, 86–7, 90–107, 118, 122, 124–5, 128, 130–1, 138
North Korea 1, 3, 131, 133–4, 140

Oceania 69
OECD 90
Offshore industries 76, 105–6, 122, 127
Oil: *see* natural resources
Open coastal cities 80, 113
Open-door policy 32, 75, 80, 82, 100, 109, 112–7, 138
Outer Islands 15, 30
Overseas Chinese (*Huaqiao*) 4, 40, 43–56, 64, 75: *see also* Chinese diaspora

Pampangga 53
Papua New Guinea 1, 38
Paracel 132
Pax Nipponica 135
Penang 25
Periodism 58
Perestroika 81
Pingshuo 108
Plantations 27, 45–6, 126
Pollution 10, 93
Polynesia 24, 38
Population 1, 4, 7–8, 14–22, 43, 95, 98, 101, 103, 106, 108, 110, 120–6
Population control 17, 29, 93, 96, 119

Population density 15, 27–8, 46, 91, 98, 100
Population growth 8, 16–18, 80, 94, 98, 108
Population pressure 15
Portugal 25, 133: *see also* colonialism
Poverty 3, 10, 26, 28–9, 45, 66, 84, 94, 107, 119–20, 123, 125
Pribumi 53
Privatisation 61
Procurement and distribution 59, 62, 138
Protectionism 72–5, 97, 102
Protestantism 40
Proto-Malays 38
Pudong 114
Pusan 100

Quality of life 8–9

Race: *see* ethnicity
Ramos 122
Red River
Refugees 10
Religion 21, 36–7, 41, 46, 51, 53, 91, 133, 136
    Buddhism 40, 45, 91, 121
    Calvinism 91
    Catholicism 17, 41, 44, 133
    Christianity 39–40, 91, 121
    Confucianism 36, 39–41, 48–50, 57, 91, 96
    Islam 21, 24, 39, 53, 121, 126, 133, 136
    Shinto 91
Remittances 122
Resources (natural) 4, 11–14, 18, 22, 27, 91–2, 94–5, 98, 100, 103, 106–8, 116, 118, 120–4, 126, 130: *see also* minerals
    (human) 4, 14–22, 92, 96, 100, 108, 116, 120–3: *see also* population
Responsibility system 80, 110
Retailing 27, 52, 59–60, 62
Rice 27–8, 30, 47, 80, 85, 93, 104, 119, 126
Rivers 11, 23

Sabah 123, 125, 133
Saigon 28, 119
Samurai 37, 91

SAR (Special Administrative Region) 99
Sarawak 47, 123, 125, 133
SCCCI (Singapore Chinese Chamber of Commerce and Industry) 48
Self-sufficiency 57, 59, 62, 72
Seoul 100
SEZ (Special Economic Zone) 80, 109, 113
Shandong 113
Shanghai 10, 51, 98, 105, 113–4
Shenzhen 99
Shipbuilding 98, 101, 104
Siberia 75
Smallholdings 27–8, 30, 101, 126
South Asia 8, 10, 67, 69
Soviet Union 74, 81, 102, 115, 117, 119, 131–2, 140
Spain 25, 45–6, 121
Spratly 132
Sri Lanka 96
Subic Naval Base 132
Subsidies 73, 93, 111
Suharto 52, 127
Sukarno 52, 54
Sultan of Brunei 126
Sulu 24, 39
Sumatra 15, 97, 132
Sung 45
Suzhou 51

Taichung 105
Taipei 105–6
Tamils 45
Technology 14, 20, 37, 40, 45–7, 66, 83, 90–4, 97, 102–3, 105, 108–9, 113, 115–7, 130
Teochew 44, 95
Textiles 98–9, 103–4
Thailand 1, 11, 96, 131
Tiananmen Square 60–1, 75, 99, 109
Timor (East) 55, 127, 133
Tokugawa period 90–1
Tourism 96, 99
Trade 4, 14, 23, 29, 32, 39, 41, 45–6, 48, 56–64, 98–100, 102–3, 113, 116, 118, 127, 129, 132, 134, 136, 138, 140
    (Domestic) 4, 23–4, 26–7, 32, 47, 56–64
    (International) 4, 23, 26–7, 32, 39, 56, 65–78, 82, 86, 97, 104–6, 131
    (Private) 59–64

Trade balance (surplus or deficit) 72,
74, 77, 92, 98, 105, 116
Transport 29, 33, 51, 90–1, 105, 108,
125

United Nations 6, 101, 106
United States 1, 4–5, 25–7, 31, 45, 65,
69–70, 72, 77–8, 92–3, 97, 101–2,
105–6, 119–22, 129–37
Urbanisation 10, 15–17, 46

Waray 53
West Pacific Rim (definitions) 1–3
Wholesaling 62

Work ethic 21, 35, 37, 48, 98
World Bank 6–8, 31, 113, 125

Xiamen, 113
Xianjiang 133
Xinyong 49

Yangtze River 110, 114

Zaibatsu 38, 91, 101
Zambia 8
Zhugong 52–3